Ergonomics and Appropriate Technologies

NIPA® GENX ELECTRONIC RESOURCES & SOLUTIONS P. LTD.
New Delhi-110 034

About the Authors

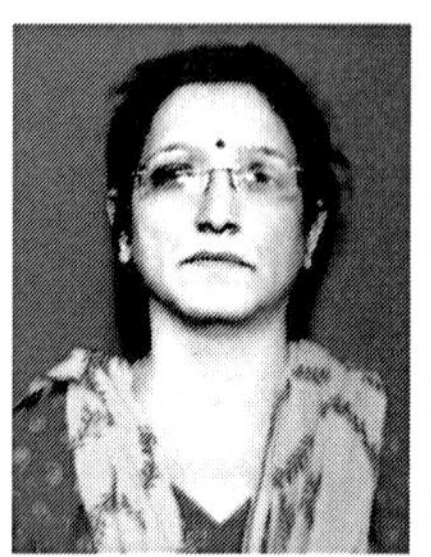

Dr. Aditi Vats is Professor & HOD in the Department of Family Resource Management, College of Home Science, GBPUAT Pantnagar. She holds Ph. D degree in Home Science. Her field of specialization is Women Empowerment, Entrepreneurship and Gender Development, Event and décor Management and basic Ergonomics. She has guided 20 research students in M.Sc and Ph.D. She has to her credit 70 research papers in international and national journals,15 popular articles,2 manuals , 2 books, course writer in 2 online courses, developed e-Content of 3 courses, attended 20 seminars and conferences and launched online certificate course under NAHEP project. Received Distinguished Scientist Award – 2019 in 3rd International Conference on Global Initiatives in Agricultural and Applied Sciences for Eco-friendly Environment in the field of Family Resource Management.

Dr. Chhaya Shukla, (Ex-Member, Uttarakhand Public Service Commission) Associate Professor, Department of Family Resource Management, College of Home Science, GBPUAT Pantnagar, holds Ph. D degree in Home Science. She is recipient of ICAR Junior Research Fellowship and ASRB-NET qualifies. Her area of specialization includes Women Entrepreneurship & Empowerment, Event and décor Management, Consumer education, housing & space management and basic Ergonomics. She has guided 05 M.Sc. research students. She has to her credit 15 research papers in international and national journals,08 popular articles, 3 chapters in books, 24 syllabus moderations, attended 11 seminars and conferences and 7 trainings and organizes 11 trainings/ workshops on various topics of Resource Management and Consumer Education.

Dr. Sandhya Rani is Assistant Professor in the Department of Family Resource Management, College of Home Science, GBPUAT Pantnagar. She holds Ph. D degree in Family Resource Management. She is also recipient of UGC-NET and Best Poster Presentation Award from reputed society for her work. She has so far published 12 research papers in the journals of national and international repute and 3 chapters in books. She has expertise in the field of Family Resource Management.

Ergonomics and Appropriate Technologies

Aditi Vats
Professor & HOD
Department of Family Resource Management
College of Home Science
GBPUAT, Pantnagar, Uttarakhand

Chhaya Shukla
(Ex-Member, Uttarakhand Public Service Commission)
Associate Professor
Department of Family Resource Management
College of Home Science
GBPUAT Pantnagar, Uttarakhand

Sandhya Rani
Assistant Professor
Department of Family Resource Management
College of Home Science
GBPUAT, Pantnagar, Uttarakhand

NIPA® GENX ELECTRONIC RESOURCES & SOLUTIONS P. LTD.
New Delhi-110 034

NIPA® GENX ELECTRONIC RESOURCES & SOLUTIONS P. LTD.

101,103, Vikas Surya Plaza, CU Block
L.S.C.Market, Pitam Pura, New Delhi-110 034
Ph : +91 11 27341616, 27341717, 27341718
E-mail: newindiapublishingagency@gmail.com
www: www.nipabooks.com

For customer assistance, please contact
Phone: + 91-11-27 34 17 17
Fax: + 91-11- 27 34 16 16
E-Mail: feedbacks@nipabooks.com

ISBN: 978-81-19103-12-6

Composed and Designed by NIPA®.

Dean College of Community Science
G.B. Pant University of Agriculture & Technology
Pantnagar – 263145, (U.S. Nagar)
Ph. 05944-233637

Dr. Alka Goel
(Dean, College of Community Science)

Foreword

The manual on Ergonomics & Appropriate Technology has been developed to furnish excellent practical experience for graduate students of College of Community Science/Home Science with the aim to improve the performance of tasks in work and of nearby environment. Ergonomics is an integral part of designing, manufacturing and use, the graduates need to be aware about the study of anthropometry, posture, repetitive, motion, workplace design thus understanding of ergonomics is very pertinent as it effect the end user's work execution and health aspects.

Embracing the concepts of designing of equipment, task performance for human comfort and satisfaction as well as optimum performance, this manual clearly defines objective of the ergonomics required in day to day life. Detailed updated techniques of the work, worker and workplace related to practical life situation have been covered adequately in this manual on

"Ergonomics & Appropriate Technologies"

I am extremely delighted to congratulate my faculty members of preparing an updated practical manual for knowledge up graduation and skill enhancement of the college graduates

Dr. Alka Goel
(Dean, College of Community Sceince)

Preface

Ergonomics is *"**an applied science concerned with designing and arranging things people use, so that the people and things interact most efficiently and safely;**"* as defined by Merriam-Webster. Ergonomics is a science concerned with the 'fit' between people and their work. The name 'ergonomics' comes from the Greek words 'ergon', which means work and 'nomos' which means law.

The goal of ergonomics which is the scientific study of people at work is to prevent soft tissue injuries and musculoskeletal disorders (MSDs) caused by sudden or sustained exposure to force, vibration, repetitive motion, and awkward posture. In all it can be said that ergonomics is a field of study that attempts to ***reduce strain, fatigue, and injuries by improving product design and workspace arrangement. The goal is to provide a comfortable, relaxed posture so as to increase efficiency and productivity of the worker at workplace.***

Dimensions of Ergonomics

- **Physical Ergonomics**
- **Cognitive Ergonomics**
- **Organizational Ergonomics**

How Does Ergonomics Work?

Ergonomics is concerned with the design of space in which people carry out work, it establishes relation of man to environment in which an individual works and its applications related to anatomical, physiological, psychological and engineering knowledge to the problems involved. Ergonomists work across and use data and techniques from several different disciplines, such as:

- **Anthropometry:** Body sizes, shapes, variations.
- **Biomechanics:** Muscles, levers, forces, strength.
- **Environmental physics:** Light, heat, noise, radiation, cold, vision, and so on.
- **Applied psychology:** Learning, errors, differences.
- **Social psychology:** Group behaviour, communication.

Ergonomics and Technology

Beyond PPE, technology also improves safety through ergonomics. Sit-to-stand desks and abundant seating options provide basic examples of ergonomic improvements that help prevent injuries from improper support. There are five aspects of ergonomics: **safety, comfort, ease of use, productivity/performance, and aesthetics**. For this we have to follow ergonomic principles.

Benefits of Ergonomics

People who work in ergonomic workplaces benefit from improved health. The effect of ergonomics starts within the cardiovascular system and spreads to other areas. Your heart will be healthier than it would be if you worked in a standard environment. Ergonomics can improve the health of employees by reducing work-related injuries, most often caused by strains and over-exertion. If ergonomics is followed properly it may lead to;

- **Improve Productivity:** Ergonomic workplaces are efficient workplaces. Ergonomic workstations combine different ideas to improve workers' ability to complete their tasks—from organizing items and supplies to integrating computer equipment and monitors. An optimized workstation allows a worker to focus on their task and not be distracted by discomfort or a lack of organization. The more focused your employees are, the higher the level of productivity they can have.
- **Improve Mental Clarity:** Reducing physical discomfort and improving your posture can improve your mental clarity and allow you to do your work more successfully. Ergonomics can also help to reduce stress and improve concentration. When you're feeling comfortable, you can focus better on your work. Ergonomics helps decrease pain, strengthen muscles, and increase blood flow. This can improve mental insight also.
- **Decrease Pains:** Ergonomics focuses on optimizing the design of the workplace, tools, and equipment to reduce strain on employees, minimize fatigue, and improve overall comfort and safety. Ergonomics can also help improve posture and reduce the risk of musculoskeletal injuries, such as back pain, by providing ergonomically designed chairs, desks, other furniture, and adjustable workstations. By creating an ergonomically designed work environment, employers can help to ensure that employees remain healthy and productive.
- **Eliminate Hazards:** Part of creating a more productive work environment is eliminating the daily hazards that can hurt your employees. Ergonomics contributes to identifying and eliminating hazards in the workplace by creating work environments that are tailored to fit the user. Additionally, ergonomics helps to create a better environment by reducing distractions,

providing suitable lighting and ventilation, and providing adjustable furniture and equipment, all of which help eliminate potential workplace hazards.

- **Improve Quality of Work:** The benefits of ergonomic workspaces range from employee well-being to the quality of work. Aches, pains, fatigue, and other problems can affect a worker. Ergonomics can eliminate those issues and help workers work. Additionally, ergonomics can help ensure that employees are using the most effective tools and equipment to do their job and that they can reach and use the tools and equipment safely.
- **Reduce Absenteeism:** Ergonomics can improve absenteeism by helping prevent workplace injuries that lead to missed work days and creating a more comfortable working environment for employees. Ergonomic practices can also reduce employee fatigue and stress, which can be physical or psychological factors contributing to absenteeism. Encourage your workers to take regular breaks, change postures frequently, and adjust their workstations to fit their body size and shape better. Finally, promote positive health behaviors by providing resources and encouraging employees to maintain a healthy diet and lifestyle.
- **Focus on Safety:** Ergonomics will create a safer work environment and increase awareness. It will remove hazards, improve workstations for less discomfort, and teach your employees to update their spaces with safety in mind.
- **Increase Employee Satisfaction:** The more you lean into the ergonomics culture, the more positive your work environment will be. Ergonomics in the workplace can help improve employee satisfaction by reducing physical and mental stress. As a result, your employees will enjoy coming to work more than they did before, affecting those around them.
- **Lower Insurance Costs:** Ergonomics in the workplace can help lower insurance costs by reducing the risk of employees suffering from work-related injuries and illnesses. By reducing the number of Workers' Compensation claims, it could be able to save on insurance premiums. Ergonomic improvements such as adjustable workstations, ergonomic chairs and keyboards, and improved lighting can help reduce employee fatigue and improve your overall working environment. In such an ergonomic work environment, your employees are less likely to become injured on the job. In addition, implementing ergonomic policies and programs can provide with greater legal protection in case of a worker injury.

The implementation of ergonomics in system design should make the system work better by eliminating aspects of system functioning that are undesirable, uncontrolled or unaccounted for. It can be concluded that Ergonomics is the human–machine system. It is a system which includes a set of elements, the relations between these elements and the boundary around them. Most systems consist of people and machines and perform a function to produce some form of output. Inputs are received in the form of matter, energy and information. For ergonomics, the human is part of the system and must be fully integrated into it at the design stage. Human requirements are therefore system requirements, rather than secondary considerations and can be stated the ergonomics include equipments that is usable and safe, tasks that are compatible with people's expectations, limitations and training, an environment that is comfortable and appropriate for the task and finally a system of work organization that recognizes people's social and economic needs.

Pantnagar **Authors**

Contents

1

Introduction To Ergonomics

Objective: to study about ergonomics and its applications.

What is Ergonomics?

The International Ergonomics Association (IEA) defines ergonomics or human factors as the scientific discipline concerned with the understanding of interactions among humans and other elements of a system, and the profession that applies theory, principles, data and methods to design to optimize human well-being and overall system performance.

The word ergonomics comes from the Greek word "ergon" which means work and "nomos" which means laws. It's essentially the "laws of work" or "science of work". It is a discipline that studies the interactions between human activity and the components of this activity (tasks, tools, methods, work environment, etc.) to develop systems that allow people to work in conditions of optimal efficiency, safety and comfort.

Ergonomics is the science of fitting the jobs to the people who work in them. It designs a job to fit the worker so the work is safer and more efficient. It concerned with the "fit" between people and their technological tools and environments. It takes account of the user's capabilities and limitations in seeking to ensure that tasks, equipments, information and environment suit each other. To assess the fit between a person and the used technology, ergonomist consider the job (activity) being done and demands on the user, the equipment used (its size, shape, and how appropriate it is for the task) and the information used (how it is presented, accessed and changed).

The process of ergonomic evaluation focuses on making the work task, work station and work environment fit the individual work rather than the other way around. Implementing ergonomics solutions can make employees more comfortable and increase productivity. By taking account of physical and psychological capabilities and limitations of humans, ergonomics can help to avoid unsafe, unhealthy, uncomfortable or inefficient situations at work. Ergonomists work across and use data and techniques from several different disciplines, such as anthropometry, biomechanics, Environmental physics, engineering, physiology, psychology etc. Ergonomics aims to create safe, comfortable and productive workspaces by bringing human abilities and limitations into the design of a workspace, including the individual's body size, strength, skill, speed, sensory abilities (vision, hearing), and even attitudes.

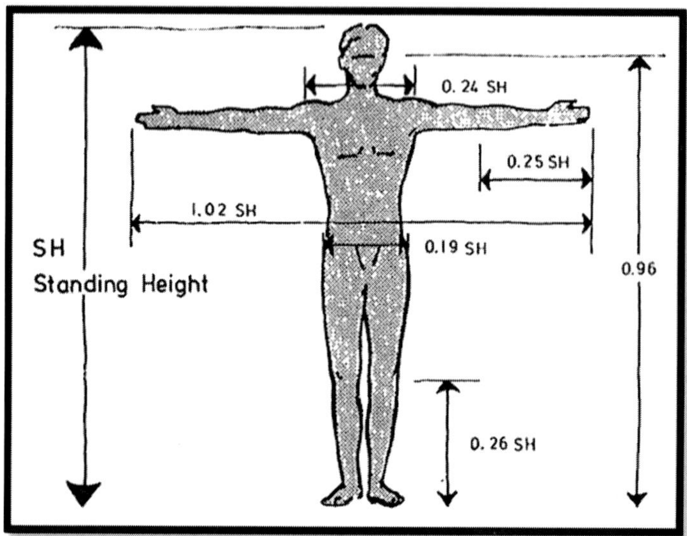

How Does Ergonomics is Work?

Ergonomics is concerned with the design of space in which people carry out work, it establishes relation of man to environment in which he works and the applications of anatomical, physiological, psychological and engineering knowledge to the problems involved. Ergonomists work across and use data and techniques from several different disciplines, such as:

- **Anthropometry:** Body sizes, shapes, variations.
- **Biomechanics:** Muscles, levers, forces, strength.
- **Environmental physics:** Light, heat, noise, radiation, cold, vision, and so on.
- **Applied psychology:** Learning, errors, differences.
- **Social psychology:** Group behaviour, communication.

Ergonomics is used in all field of designing. One of the aspects where finds an importance place is designing life space for family living.

Application of Ergonomic Principles

- All the structural features in the house should be according to its intended purpose/function and user's anthropometry.
- There should not be any visual discomfort in any part of the house. Every work surface should be well lighted, ventilated without any glare and shadow.
- Every item accessed frequently should be within reachable distance to avoid over-stretching.
- Adjust work surface height so that it is at or below elbow height for most job tasks.
- Adjust chair or work stool height according to leg length and the height of the work surface.
- Allow space so that legs can be outstretched.
- Handgrips and tools used in work should fit into hand size.
- Workstation should be arranged to avoid over-stretching and unnecessary twisting.
- When sitting, sit upright with ankles, knees, hips, and elbows positioned at ninety degrees with shoulders relaxed.
- Select an adjustable chair with a stable base, adequate padding, good low-back support, proper back and hip support, and the seat doesn't hit the back of the knees.
- When doing housework try to keep pathways clear to avoid unnecessary twisting or stooping while mopping or sweeping.
- Whether at home or at work you can avoid serious, costly injuries and improve stamina with use of these simple ergonomic principles.

Exercise-1

Answer the following questions:

1. How does ergonomics work in the following areas?

Anthropometry

Environmental physics

2. Application of ergonomics principles in designing of household equipments:

Conclusion:

Remarks:

Signature

2

Domains of Ergonomics

OBJECTIVE: To study about domains of ergonomics.

Domains of Ergonomics

IEA divides ergonomics broadly into three domains:

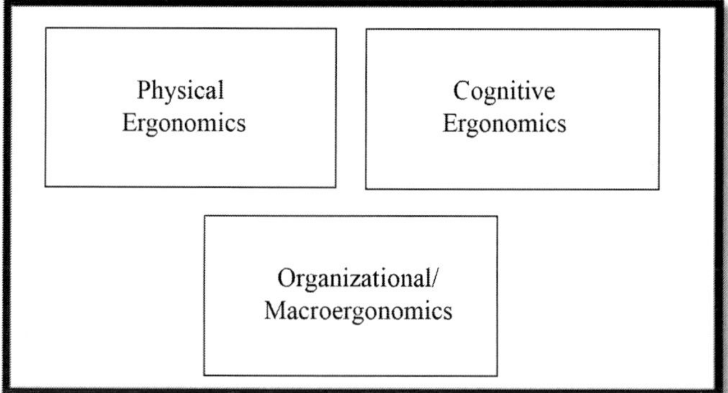

Each of these differently affects how people interact with their environment, contributing to how effective they are in their work.

Physical Ergonomics

Physical ergonomics is concerned with human anatomical, anthropometric, physiological and biomechanical characteristics as they relate to physical activity. This is the ergonomics domain we are most concerned with in the workplace, and most of the content on this site is very much focused on workplace ergonomics. Physical ergonomic principles have been widely used in the design of both consumer and industrial products for optimizing performance and to preventing / treating work-related disorders by reducing the mechanisms behind mechanically induced acute and chronic musculoskeletal injuries / disorders. Risk factors such as localized mechanical pressures, force and posture in a sedentary office environment lead to injuries attributed to an occupational environment. Physical ergonomics is important to those diagnosed with physiological ailments or disorders such as arthritis (both chronic and temporary) or carpal tunnel syndrome. Pressure that is insignificant or imperceptible to those unaffected by these disorders may be very painful, or render a device unusable, for those who are. Many ergonomically designed products are also used or recommended to treat or prevent such disorders, and to treat pressure-related chronic pain.

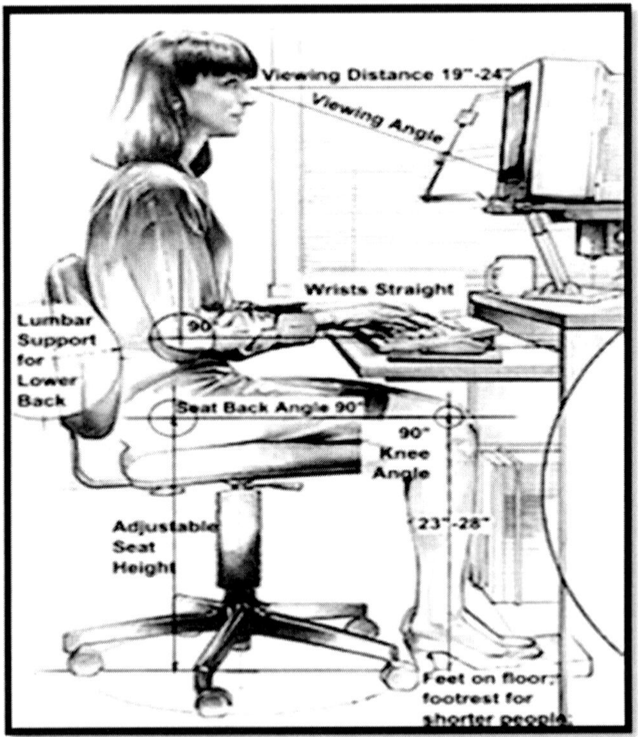

Cognitive Ergonomics

Cognitive ergonomics is concerned with mental processes, such as perception, emotion, memory, reasoning, and motor response, as they affect interactions among humans and other elements of a system. Relevant topics include mental workload, decision-making, skilled performance, human reliability, work stress and training as these may relate to human–system and human–computer interaction design. Epidemiological studies show a correlation between the time one spends sedentary and their cognitive function such as lowered mood and depression.

Organizational Ergonomics

Organizational ergonomics is concerned with the optimization of socio-technical systems, including their organizational structures, policies, and processes. It also includes human communication successes or failures in adaptation to other system elements, resource management, work design, work systems, design of working times, teamwork, participatory ergonomics, community ergonomics, cooperative work, new work programs, virtual organizations, remote work, and quality management. Safety culture within an organization of engineers and technicians has been linked to engineering safety with cultural dimensions including power distance and ambiguity tolerance. Low power distance has been shown to be more conducive to a safety culture. Organizations with cultures of concealment or lack of empathy have been shown to have poor safety culture.

Workplace Ergonomics

The science of fitting workplace conditions and job demands to the capabilities of the working population. Ergonomics is an approach or solution to deal with a number of problems—among them are work-related musculoskeletal disorders. At its core, workplace ergonomics is really about building a better workplace. When jobs are designed to match the capabilities of people, it results in better work being produced and a better experience for the person doing it. The ergonomics improvement process systematically identifies ergonomic hazards and puts in place engineering and administrative control measures to quantifiably reduce risk factors. Ergonomics is important because when you're doing a job and your body is stressed by an awkward posture, extreme temperature, or repeated movement your musculoskeletal system is affected and it can be reduced if proper ergonomic principles are applied.

Ergonomics Process

Step by step procedure can be applied to reduce risk factors in doing task at either in office or at home. The steps to be followed are

- **Assess Risk**: Conducting an ergonomic assessment is a foundational element of the ergonomics process. Ergonomic improvement efforts will never get off the ground without being able to effectively assess jobs in your workplace for musculoskeletal disorder (MSD) risk factors.
- **Plan Improvements**: The core goal of the ergonomics process is to make changes to your workplace that reduces risk. Making changes at scale requires a significant planning effort that includes prioritizing jobs to be improved, identifying effective improvement ideas, and cost-justifying the improvement projects.
- **Measure Progress**: Measurement is an important component of any successful continuous improvement process. High performing ergonomics programs are constantly measured using both leading and lagging indicators.
- **Scale Solutions**: By establishing a common set of tools to train your workforce, assess risk, plan improvements, measure progress, and design new work processes, you'll be able to scale ergonomics best practices throughout your organization.
- Habits formed in our daily activities like staring at the computer, cradling the phone, slouching in a chair or even just carrying groceries can create painful body conditions. These conditions can be helped or prevented by incorporating good ergonomic techniques into daily routines and work sites.

Exercise-1

Write and draw a situation where you felt pain and discomfort while doing work. What changes would you recommend for that situation?

Conclusion:

Remarks:

Signature

3

Principles of Ergonomics

OBJECTIVE: To understand about principles of ergonomics.

Fundamental Principles of Ergonomics

There are 8 fundamental Ergonomic Principles for better work performance which help you identify ergonomic risk factors and maintain your stellar safety record.

- Maintain neutral posture
- Work in power zone
- Allow for movement and stretching
- Reduce excessive force
- Reduce excessive motion
- Minimize contact stress
- Reduce excessive vibration
- Provide adequate lighting

Principle 1: Maintain Neutral Posture

Whether you work sitting or standing and moving throughout the day, it's important to keep your body in a neutral posture. Neutral postures are postures where the body is aligned and balanced while either sitting or standing, placing minimal stress on the body and keeping joints aligned. Neutral postures minimize the stress applied to muscles, tendons, nerves and bones and allows for maximum control and force production. The opposite of a neutral posture is an "awkward posture." Awkward postures move away from the neutral posture toward the extremes in range of motion. This puts more stress on the worker's musculoskeletal system, is a contributing risk factor for Musculoskeletal Disorders (MSDs) and should be avoided. There are several areas of the body that this principle commonly affects. Following are examples of Neutral vs. Awkward postures for the wrist, elbow, shoulder and back. When you put on your "ergo eyes", you'll immediately begin to notice when workers are in awkward postures and when they are maintaining a neutral posture.

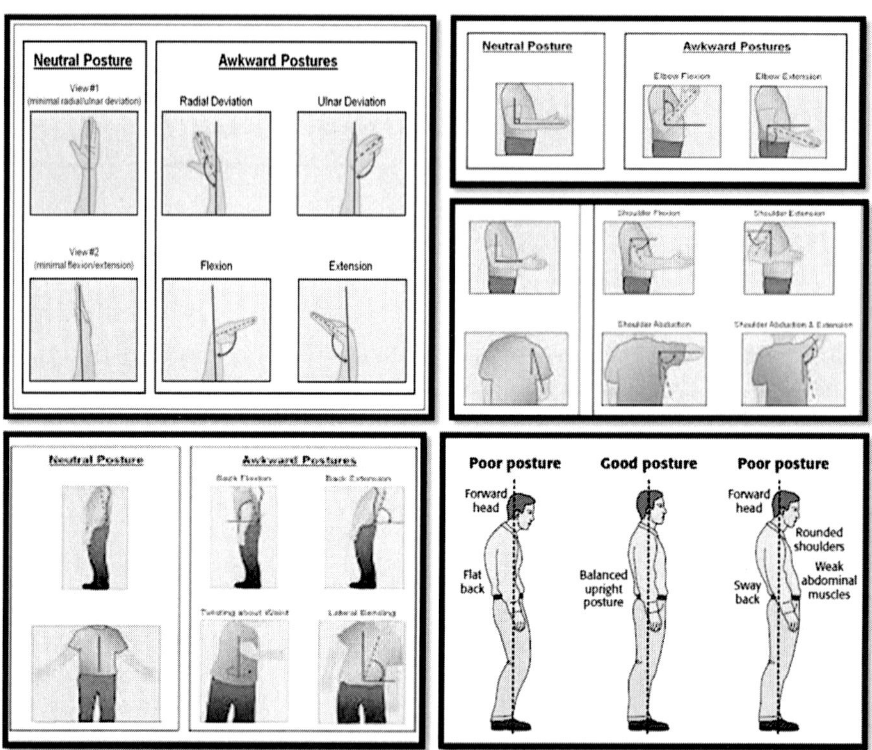

Principle 2: Work in the Power Zone

The power zone for lifting is close to the body, between mid-thigh and mid-chest height. This zone is where the arms and back can lift the most with the least amount of effort. This can also be called the "hand shake zone" or "comfort zone." The principle here is that if you can "shake hands with your work", you are minimizing excessive reach and maintaining a neutral posture.

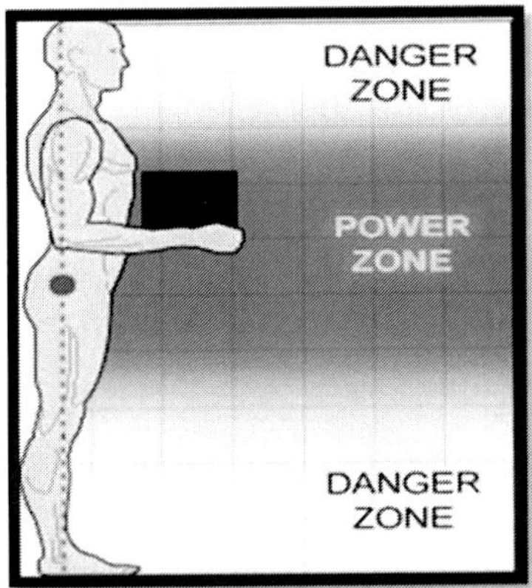

Working from the power/comfort/ handshake zone ensures that you are working from proper heights and reaches, which reduces MSD risk factors and allows for more efficient and pain-free work. Now when you notice workers who are working with extended reaches and at improper heights, you'll know they are outside their comfort zone and risk factors are present.

Principle 3: Allow for Movement and Stretching

The musculoskeletal system is often referred to as the human body's movement system, and it is designed to move. Working for long periods of time in a static position will cause your body to fatigue and discomfort. This is what is known as static load.

For example

- Raise your hands over your head for the next 30 minutes
- Remain standing in the same position for the next 8 hours
- Write with a pencil for 60 minutes straight

If you do these things, you will experience static load. The first few seconds or minutes don't seem too bad, but the cumulative effect of holding these seemingly stress-free positions over time will cause fatigue and discomfort. **Now, what is the first thing you will naturally do once you when you are finished with these tasks?**

- You will stretch.
- You'll stretch out your shoulders and back.
- You'll stretch out your legs and maybe do some squats.
- You'll stretch out your fingers and wrist.

Warm up and stretch before activities that are repetitive, static or prolonged. Stretch out your shoulders and back, legs, fingers and wrist. Stretching reduces fatigue, improves muscular balance and posture and improves muscle coordination. Everyone is an athlete in life, so you need to prepare your body for work by warming up to improve performance and lower injury risk. A warm-up stretching is a great way to prepare your body for work. It is also beneficial to take periodic stretch breaks over the course of your work day to get your blood moving and restore your energy.

Principle 4: Reduce Excessive Force

Excessive force is one of the primary ergonomic risk factors. Many work tasks require high force loads on the human body. Muscle effort increases in response to high force requirements which increases fatigue and risk of an MSD. There are numerous conditions that affect force, but the idea is to recognize when a job or task requires excessive force and then find ways to reduce that force. Eliminating excessive force requirements will reduce worker fatigue and the risk of MSD formation in most workers. Using mechanical assists, counter balance systems, adjustable height lift tables and workstations, powered equipment and ergonomic tools will reduce work effort and muscle exertions.

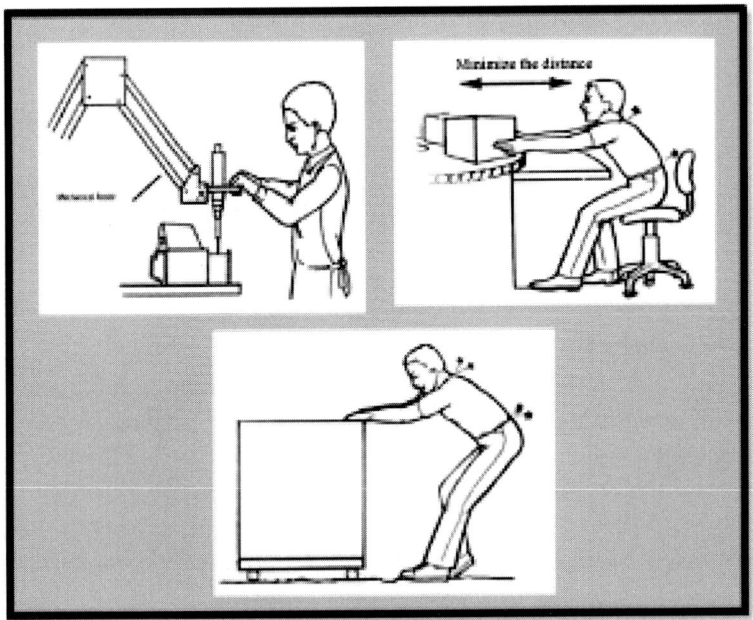

Imagine a time when you had to use your full body weight to move an object. That's what this principle is referring to. Pushing, pulling, and lifting heavy items can strain your joints, creating the potential for fatigue or injury. Rather than using excessive force, try to identify tools or strategies that can lighten the load you need to move. It may be as simple as using a cart or hoist to move heavy items, or you may need to make changes to your workflow to lessen the distance you need to travel or the number of items you need to transport.

Principle 5: Reduce Excessive Motions

Repetitive motion is another one of the primary ergonomic risk factors. Many work tasks and cycles are repetitive in nature, and are frequently controlled by hourly or daily production targets and work processes. High task repetition, when combined with other risks factors such high force and/or awkward postures, can contribute to the formation of MSD. A job is considered highly repetitive if the cycle time is 30 seconds or less. Excessive or unnecessary motions should be reduced if at all possible. In situations where this is not possible, it is important to eliminate excessive force requirements and awkward postures. Other control methods to consider are Job enlargement, job rotation and counteractive stretch breaks.

Manual repetition can cause overuse injuries, so it's important to think about the motions you repeat throughout the day and find ways to reduce excessive motion. For example, rather than twisting your wrist to turn a screwdriver, is it possible to replace the screwdriver with a drill that will take the manual motion out of the task altogether? You can also look for opportunities to change your position or the layout of your work area to help you work more ergonomically.

Principle 6: Minimize Contact Stress

According to OSHA, contact stress results from continuous contact or rubbing between hard or sharp objects/surfaces and sensitive body tissue, such as soft tissue of the fingers, palms, thighs and feet. This contact creates localized pressure for a small area of the body, which can inhibit blood, nerve function, or movement of tendons and muscles. Examples of contact stress include resting wrists on the sharp edge of a desk or workstation while performing tasks, pressing of tool handles into the palms, especially when they cannot be put down, tasks that require hand hammering, and sitting without adequate space for the knees.

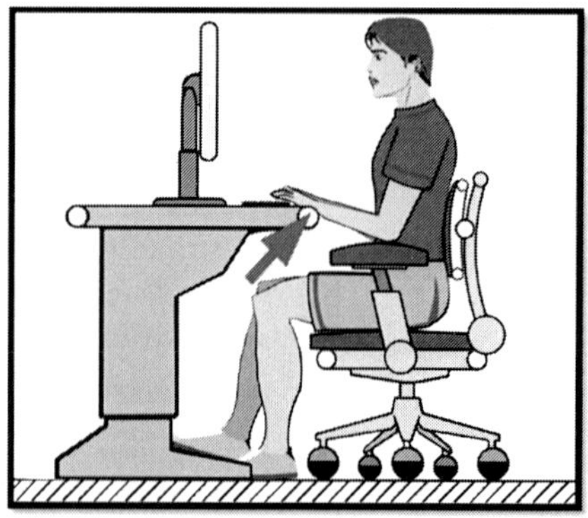

The image shows the wrong position of the wrist and keyboard. An extended time in this position leads to pressure on the carpal tunnel and median nerve which may result in carpal tunnel syndrome.

Contact stress occurs when a tool or surface routinely comes into contact with the same area of your body. Sometimes called pressure points, these areas of contact can cause discomfort. One example of contact stress is when you routinely squeeze a tool, like pliers, or hold a heavy item, like a nail gun, that puts pressure on an area of your hand. Adding padding, wearing gloves, or choosing tools with a cushioned grip can help. You may also consider adding anti-fatigue mats to standing surfaces to help alleviate contact stress on your heels.

Principle 7: Reduce Excessive Vibration

Multiple studies have shown that regular and frequent exposure to vibration can lead to permanent adverse health effects, which are most likely to occur when contact with a vibrating tool or work process is a regular and significant part of a person's job. Hand-arm vibration can cause a range of conditions collectively known as hand-arm vibration syndrome (HAVS), as well as specific diseases such as white finger or Raynaud's syndrome, carpel tunnel syndrome and tendinitis. Vibration syndrome has adverse circulatory and neural effects in the fingers. The signs and symptoms include numbness, pain, and blanching (turning pale and ashen).

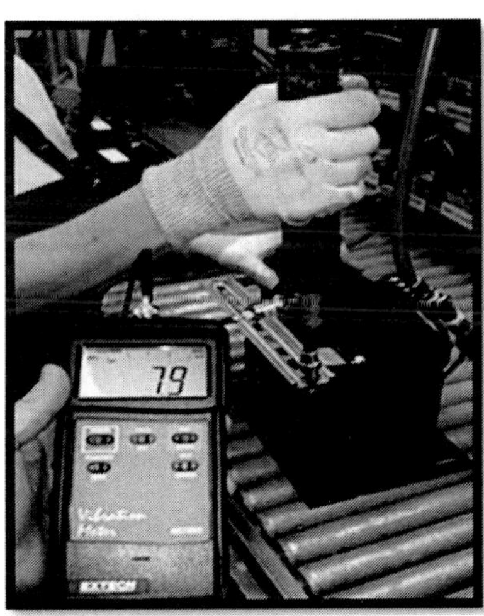

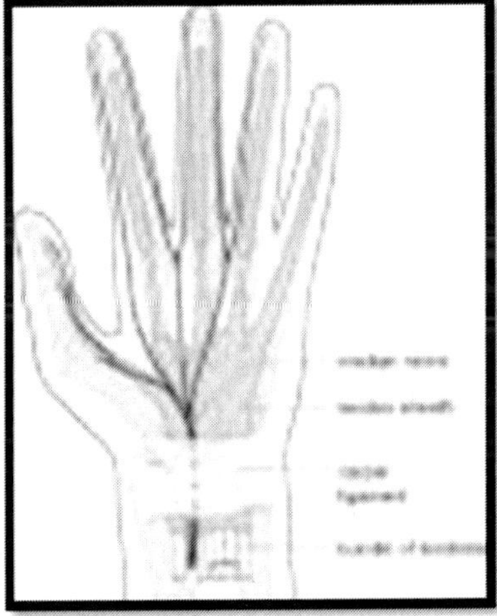

Principle 8: Provide Adequate Lighting

Poor lighting is a common problem in the workplace that can affect a worker's comfort level and performance. Too much or too little light makes work difficult – just imagine trying to do your job without sight! Dimly lit work areas and glare can cause eye fatigue and headaches and improperly lit areas put workers at greater risk for all types of injuries. Providing workers with adjustable task lighting is often a simple solution to lighting problems. At a computer workstation, take steps to control screen glare, and make sure that the monitor is not placed in front of a window or a bright background.

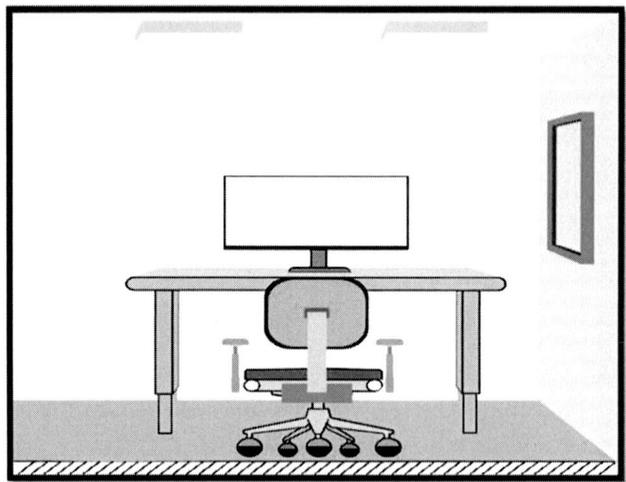

Work environments vary greatly depending on your industry and role, but some common items you may want to pay attention to are lighting, temperature, vibration, and noise. Look for ways to reduce glare or provide better lighting in low-light areas. Keep work areas at a comfortable temperature when possible, and when you're working outdoors, dress appropriately for the weather. And if tools create excessive noise or vibration, provide hearing protection and consider ways to dampen vibrating tools.

Exercise-1

Design and draw a workstation for office work on a computer system according to the principles of ergonomics. What changes would you recommend for the workstation to improve the productivity and efficiency or work?

Conclusion: ____

____

Remarks: ____

____

Signature

4

Base Materials, Insulating Material and Finishes

Objective: To identify the base materials, insulating materials, and type of finishes used in home tools and equipments.

Base Materials

The parent material to which a coating or plating is applied is known as the base material. The phrase base material is used to distinguish the overall part, numerous layers, and the original base material. Coating, electroplating, welding, and soldering all depend heavily on base materials. The following are some of the basic materials used in household tools and equipment:

- Aluminium
- Iron
- Steel
- Brass
- Copper
- Tin
- Glass
- Plastic
- Wood

Insulating Materials

Frequently used in home equipment to keep heat where it belongs and to reduce sound and noise transfer. The following are some of the insulating materials found in household items and equipment:

- Fibre Glass
- Mica

- Mineral Wood
- Plastic Foam
- Asbestos
- Rock And Mineral Wool

Finishes and Their Types

Finishes are applied to the final surface of an element as part of the construction or manufacturing process. They can provide protection from impact, water, frost, corrosion, abrasion, and other factors, as well as be beautiful. The end product determines the sort of finish on a base material. They can be used to prevent rust, make a product more appealing, improve heat conductivity, and make it easier to maintain. There are two types of finishes

- Mechanical finishes
- Applied finishes

Applied finishes are used in materials that are affected by rust and food, whereas mechanical finishes are utilised in materials that are impacted by rust and food.

Types of Mechanical Finishes

- Anodized Finish
- Silicon Finish
- High Polish Finish
- Hammered Finish
- Satin Finish

Types of Applied Finishes (two types-metallic and non-metallic)

- Metallic- Chromium, Copper, Nickel, Tin, Zinc.
- Non-Metallic- Porcelain Enamel, Synthetic Enamel, Paint, Teflon, Polyamide Finish.

Identification of the base materials, insulating materials, and type of finishes used in home tools and equipments

S. No.	Tools/ Equipment/ Appliances	Pictures	Base Materials	Insulating Materials	Finishes Used
1.	Saucepans and sauce-pots		Aluminum, Stainless Steel, Cast Iron, Brass, Copper	Plastic	Copper, Ceramic, Marble, , Porcelain Enamel, Teflon
2.	Double boiler		Stainless Steel	Plastic	
3.	Fry pans		Aluminum/Anodized Aluminum, Copper, Stainless Steel.	Plastic	Nonstick, Teflon,

4.	Griddles		A36 Steel (low carbon steel), Stainless Steel, Chrome Steel, Aluminum, Carbon Steel.	Teflon	Teflon
5.	Wok		Iron, Aluminum, Steel, Brass, Copper	Plastic	Nonstick, Teflon, Copper
6.	Dutch oven		Cast Iron, Cast Aluminum, Ceramic	Plastic	Enameled ware

7.	Cookie/Baking sheet		Aluminum, Sheet Iron, Tinned Steel	-	
8.	Cake/Pizza/Muffin/Loaf Pans		Aluminum, Oven Glass, Tinned Steel, Iron	-	Silicon, Enameled ware
9	Pie plates/pans		Aluminum, Tin, Steel Glass,	-	Enameled steel
10.	Casseroles		Glass, Ceramic, Stain-less Steel, Stands of Nickel/Chromium	Ceramic	Porcelain enamel

11.	Roasting pan		Aluminum, Iron, Steel	-	Enameled ware
12	Measuring Cups/Jugs/ Spoons		Tin, Plastic, Stainless Steel, Aluminum, Heatproof Glass	Plastic materials	
13.	Spoons	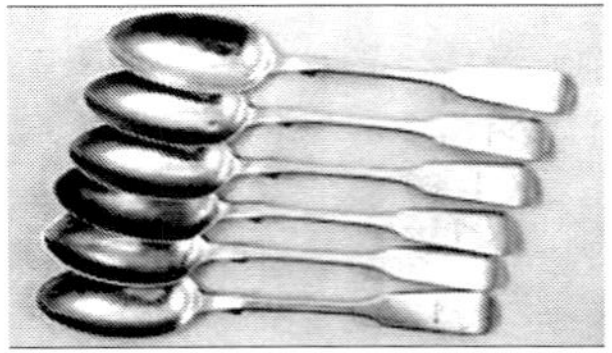	Aluminum, Tinned Steel, Stainless Steel, Iron, Brass, Copper, Plastic, Wood	Plastic, Wood	enameled ware plated silver
14.	Beaters	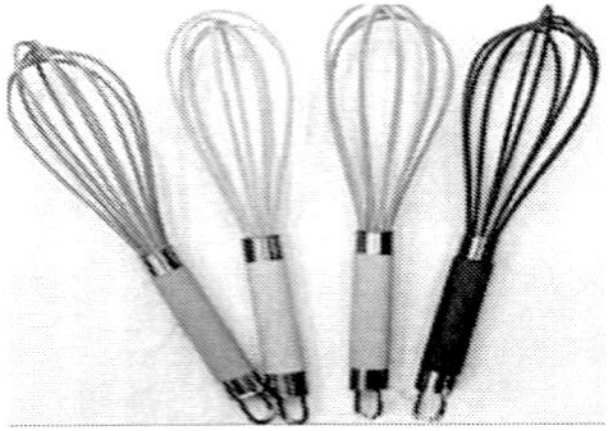	Aluminum, Stainless Steel, Plastic	Plastic	

15.	Potato mashers		Iron, Steel, Aluminum with Plastic, Metal Wooden Handles	Wood, Plastic	
16.	Knives		High Carbon Steel, Stainless Steel, Chromium, Vanadium Steel, Sheet Iron with plastic and wooden handles	Wood, Plastic	Chrome Plated,
17.	Graters/Peelers		Aluminum, Stainless Steel, Tinned Steel	Plastic	
18.	Kitchen Shears	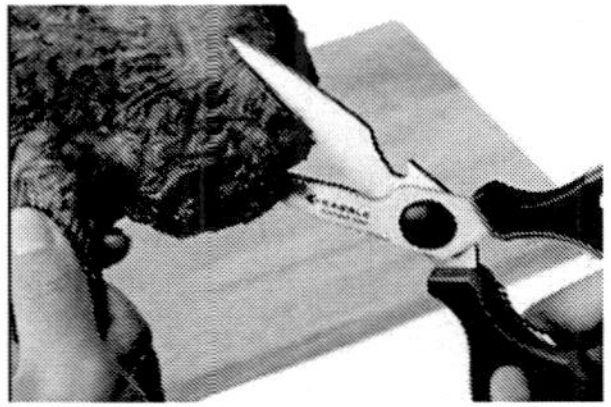	Stainless steel	Plastic or rubber coated handle	

19.	Cutting Boards		Wood, Plastic, Glass, Steel, Marble,	Wood, Plastic	
20.	Spatulas/Ladles/ Skimmers		Wood, Stainless Steel, Aluminum, Brass, Iron, Plastic, Rubber	Wood, plastic, rubber	
21.	Strainers/Colanders		Stainless Steel, Plastic	Plastic	
22.	Drum sieves/Sifters		Wood, Plastic, Tinned/ Stainless Steel with Fine Metal, Nylon Or Cheesecloth Mesh	-	

23.	Tongs		Iron, Aluminum, Steel	-	
24.	Rolling pins		Glass, Ceramic, Silicone, Wood, Stainless Steel, Marble, And Plastic		
25.	Pressure cookers		Aluminum, Stainless Steel (Body), Plastic Materials (Handles & Knobs), Rubber or silicone (Gasket)	Bakelite (thermosetting plastic)	stamped, polished, or anodized
26.	Manual juicer		Plastic, Cast Iron, Aluminum, Stainless Steel		

27.	Electric Rice Cooker and Electric kettle		Aluminum, Stainless Steel, Plastic	Plastic	
28.	Electric/Commercial Juicers		Stainless Steel, Plastic		
29.	Cooking range		Stainless steel, ceramic, glass or porcelain, enamel or glass-ceramic	Fiberglass	Porcelain enamel, chromium
30.	Ovens		Galvanized or stainless steel, ceramic, glass	Fiber glass, rock wool, cellulose	Silicon, Porcelain enamel, acrylic enamel

31.	Microwave		Glass, Paper, Ceramic, Plastic Containers		Silicon,
32.	Halogen Ovens		Glass, Stainless steel, Plastic		
33.	Solar Cooker		Iron, Glass, Wood		
34.	Refrigerator	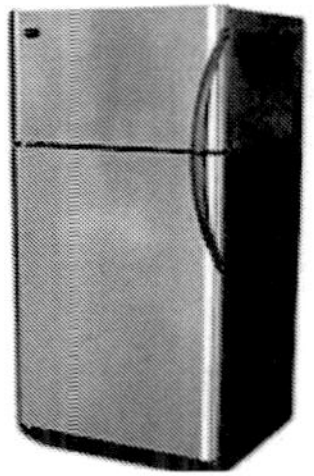	Aluminum, Cast Iron, Chrome, Copper, And Steel	Polyurethane rigid foam	Silicon, chromium

35.	Vacuum Cleaner		Combination of metals and various plastics.	Polytetrafluoroethylene (PTFE or Teflon)	
36.	Mixer Grinders		High quality plastic (outer body), stainless steel (jars) and components for motor		
37.	Coffee Makers		Plastic, steel or anodized aluminum, heat-proof glass.		Anodized

38.	Air Fryer		Plastic and metal basket		
39.	Dishwasher		Stainless Steel Or Plastic		Silicon, Porcelain enamel,
40.	Washing Machine		Porcelain-Coated, Plastic, and Stainless Steel		

41.	Iron		Iron, Plastic	Mica	
42.	Kitchen Chimney		Aluminum, stainless steel and fiber		Porcelain enamel,
43	Toasters		Plastic	Nichrome wire, Mica	

44.	Water Heaters/Geysers		Copper, Stainless steel, Thermoplastic	Nichrome wire, ribbon, or strip. Stainless steel anti-corrosion coating	Porcelain enamel, Glass Lined Vitreous Enamel Blue Silicon Enamel
45.	Air Conditioner		Plastic, copper, aluminum and stainless steel	Copper tubes and aluminum finstocks	Zinc coating corrosion-resistant powder coating spray

Exercise-1

Fill the details of at least 10 observed tools/ equipments/ appliances in the table provided below:

S. No.	Tools/ Equipments/ Appliances	Base Materials	Insulating Materials	Finishes Used
1.				
2.				
3.				
4.				
5.				
6.				
7.				
8.				
9.				
10.				
11.				
12.				
13.				
14.				
15.				

Conclusion: ______________________________

____

____

____

Remarks: ______________________________

____

____

____

____

Signature

5

Anthropometric Measurements

Objective: To study anthropometric measurements

Introduction

Anthropometry is the study of the measurement of the human body in terms of the dimensions of bone, muscle, and adipose (fat) tissue. It is a branch of anthropology dealing with measurements of human body to determine difference in groups and individuals. The word "anthropometry" is derived from the Greek word "anthropo" meaning "human" and the Greek word "metron" meaning "measure" **(Ulajaszek, 1994).** The field of anthropometry encompasses a variety of human body measurements. Weight, stature (standing height), recumbent length, skinfold thicknesses, circumferences (head, waist, limb, etc.), limb lengths, and breadths (shoulder, wrist, etc.) are examples of anthropometric measures.

It is used to describe the user or target population for a product. Answers are given in terms of the range of human body dimensions which exists in that population. It is important to designers because it is a tool that helps us to create products, interiors or exterior, equipments etc. which are suitable to most human dimensions.

Types of Anthropometric Measurements

Anthropometry deals with measurements of the human, external body dimensions in static and dynamic position. This includes measurements of body parts, their strength, speed and ranges of their motion.

- **Static Anthropometry (Structural Measurements):** Static measurements taken when the body is in a fixed or static position. These dimensions are measured and recorded by ergonomists with the body of the persons in fixed or static standardized positions. The two standard static positions are standing and sitting positions. Many physical characteristic's can be measured in these standard conditions. These measurements can have specific applications while designing personnel goods and safety appliances.

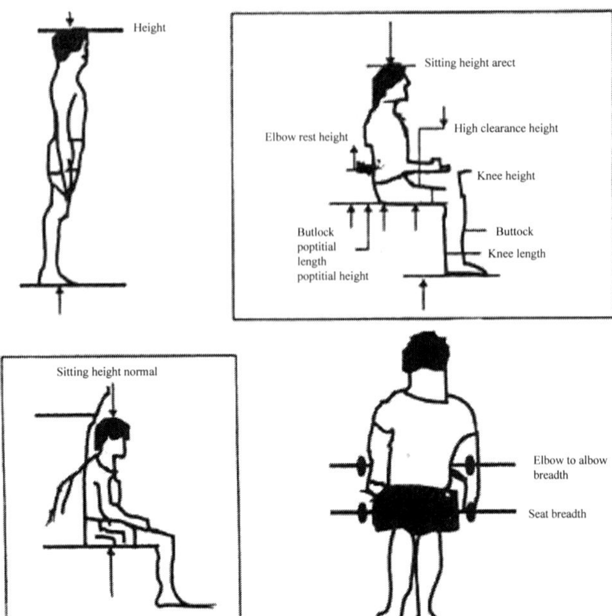

- **Dynamic Anthropometry (Functional Measurements):** Dynamic dimensions are taken under condition in which body is engaged in some physical activity. It is mainly concerned with aspects such as reach capabilities, freedom of movement of joints, body postures, visual fields and forces. Dynamic anthropometry is required more than static body measurements in the field of design. The human body is not rigid but rather always dynamic. The measurements taken of body positions which result from its movements or motions are known as functional body dimensions. Most of the times, human beings do not remain static but move their limbs in order to perform certain functions. In almost all movements the individual limbs does not move independent entry but in concert. The limiting factor in arm movement is not just the arm length but it is also affected by movements of shoulder and trunk and the function that is to be performed by the hand.

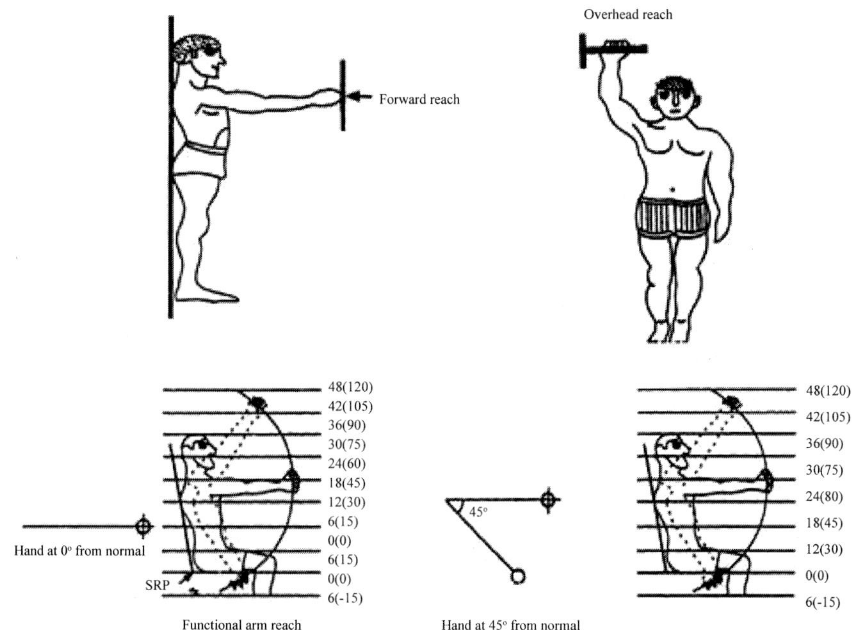

Measuring Instruments

Typical equipment list required to obtain anthropometric measurements includes:

- Weight scale
- Stadiometer
- Sliding caliper
- Spreading caliper
- Skinfold caliper
- Nonstretchable tape measure
- Measuring tape

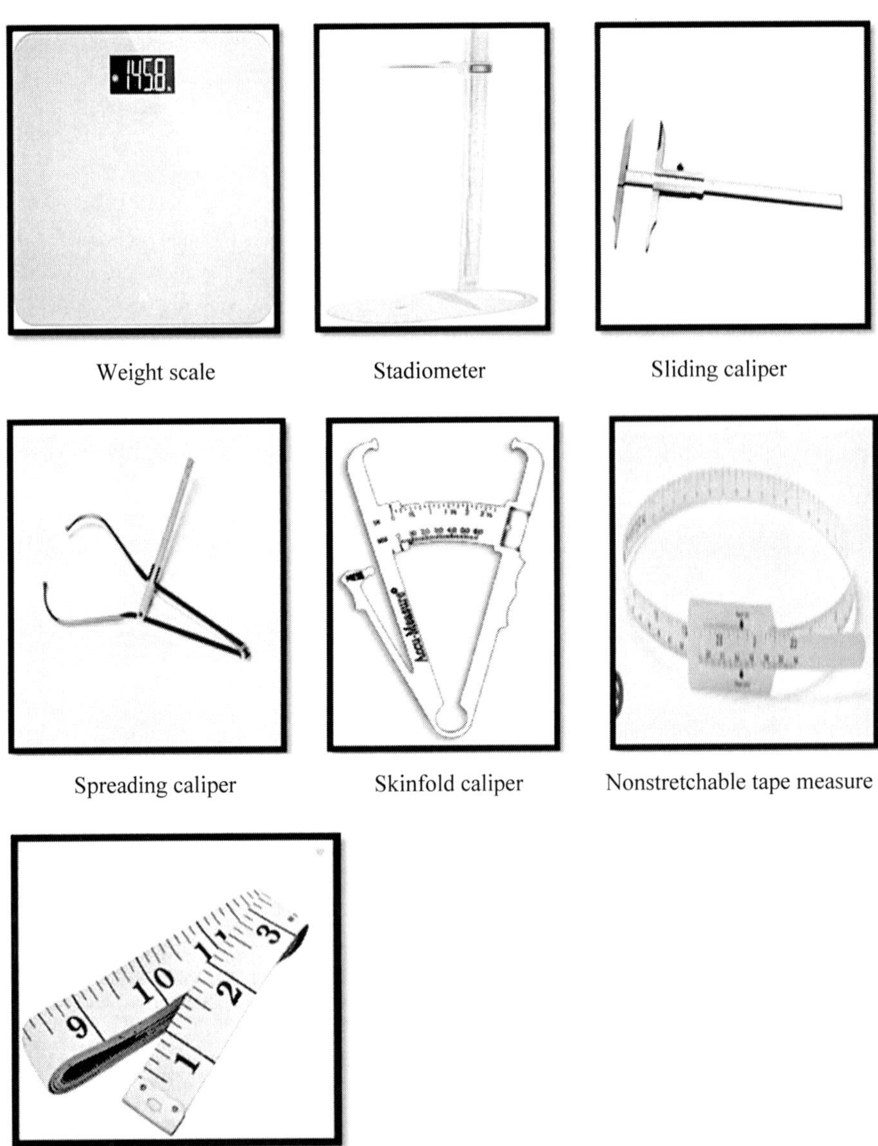

Weight scale

Stadiometer

Sliding caliper

Spreading caliper

Skinfold caliper

Nonstretchable tape measure

Measuring tape

Methods of Recording

There are three methods for collecting anthropometric measurements

- Analyzing photographs of body parts
- Directly measuring parts of body

- Statistical means (deprivation from other measurements)

Uses of Anthropometric Data

- To fit the task to the person
- To avoid physical mismatches between dimensions of equipments abd products and corresponding user.
- To design for variability in people and not for the average.

Methodology

The core elements of anthropometry are height, weight, head circumference, body mass index (BMI), body circumferences to assess for adiposity (waist, hip, and limbs), and skinfold thickness. Following methods are used to measure the height, weight and calculating the BMI.

- **Measuring Weight**
  - The subject should wear minimal, light clothing
  - Ask the subject to remove any heavy objects and take off his/her shoes.
  - Ensure the scale is on a perfectly flat surface with no obstructions.
  - Ask the subject to stand on the scale, placing feet on the footmarks pasted on the scale, and remain still until the weight appears.
  - Read and record the weight in the appropriate section of the form.
- **Measuring Height**
  - Steps of Measuring Height using a Height Measuring Scale
    - Ask the subject to remove their shoes and any headwear or hair ornaments
    - Raise the height of rod
    - Help the subject to stand if required
    - Tell the subject to stand straight
    - Lower the height rod until it rests on the person's head
    - Read the reading.
    - Raise the rod.
    - Ask the patient to come out

- Steps of Measuring Height using a Inch Tape
  - Put mark on the wall with the help of inch tape before patient arrives.
  - Help the patient to stand back against the wall and on the heels.
  - Tell the patient to stand straight.
  - Take a scale or flat board and rest on the patients head.
  - Read the reading
- Calculation of Body Mass Index
  - Body Mass Index (BMI) is an estimate of body composition that correlates an individual's Weight and Height to lean body mass. The BMI is thus an index of weight adjusted for structure. Body mass index is figured by dividing weight in kilograms and height in meter squared.

    BMI = weight in kilograms / [height in meters x height in meters]

Exercise-1

Take the following anthropocentric measurements of subjects and calculate BMI of each:

Observation Table:

S. No.	Name of Subject	Height (cm)	Weight (kg)	BMI

S. No.	Name of Subject	Height (cm)	Weight (kg)	BMI
Total Average				

Results: Average of observations (in inches):

1. Average height of the respondents :
2. Average weight of the respondents :
3. Average BMI of the respondents :

Calculation Sheet

Conclusion: ____

____

____

____

____

____

Remarks: ____

____

____

____

____

____

____

____

Signature

6

Normal and Maximum Reach Zones

Objective: To determine normal and maximum reach zones at vertical and horizontal planes.

Equipments Required: Anthropometric kit, Measuring tape, scale, marker etc.

Work Area

The rate of consumption of time and energy in task completion is affected by the work area. The work area is composed of the height, available space of the work surface, the location of the equipment, materials, and supplies being used. The work surface should be at the optimum height for the position you assume in task completion. This means if you are standing, the surface height should be slightly below your waistline. During the task completion there should be of necessity to end at the waistline or hipline to complete any actions involved in the task. When in a seated position, your work surface should not require you to reach upward nor should there be excessive bending of the neck in order to view the activity in progress. Excessive bending at the waistline, in all probability, also indicates your work surface is too low. Both feet should rest comfortably on the floor. Good posture should be observed at all times.

Closely allied to the work surface is the location of the materials, supplies, and equipment. You have two reaching zones affecting your energy consumption.

- The **maximum reach zone**, shown in **Figure-1**, is created when your arms are fully extended outward from the body to form a circle or arch. This circle represents the farthest can reach without stretching or moving another part of your boy. Any task whether seated o standing, should be performed within your maximum reach zone. Tasks that extend beyond this area require additional movement and thus necessitate further energy and time consumption.
- The **normal reach**, shown in the **Figure-2**, is formed by the circle extending outward from your body when your arms are bent at the elbows. It is within this zone that most of your work actually takes place.

The specific task undertaken is performed here. Within this circle you have the optimum view and the most effective work motions.

Fig. 1: Maximum Reach Zone

Fig. 2: Normal Reach Zone

Both of these reach zones have an effect upon your time and energy consumption. Any materials, supplies, and equipment necessary for task completion should be located within the maximum reach zone. The specific action occurs within the normal reach zone.

Although these two zones are important for task completion, they also play an important role in organizing work centers. Equipment, materials, and supplies used on a regular basis should be placed in these zones, and where possible on upper cabinets. You use less energy reaching upward than when reaching downward. Heavy equipment should also be stored in upper rather than lower cabinets, where possible.

Procedure

The dimensions for normal as well as maximum work areas for both vertical and horizontal work surfaces are required to know the space without any disturbance and work surface areas for smooth functioning. If the working area is beyond reaching power it may cause physical fatigue which may lead to physiological fatigue directly or indirectly. Therefore, to carry out work smoothly, horizontal and vertical reaches are required for work simplification.

- The maximum work area for an individual in either standing or sitting position may be determined by extending each arm from the shoulder and making a circular sweep with the shoulders as pivotal points. The over lapping areas formed by the areas of the right hand and left hand constitute the zone beyond two hands cannot work close together out moving the body.

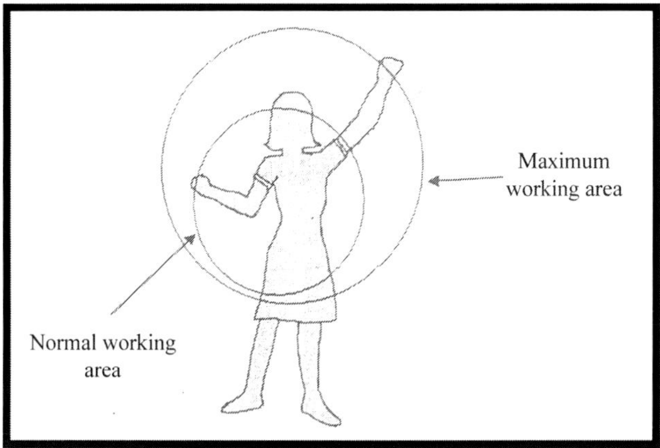

Fig. 1: Normal and Maximum Working Area in Vertical Plane

- The normal or comfortable work areas are determined in similar fashion except in the case only the fore arms are extended the upper arm begins more or less relaxed and the elbows closed to body.

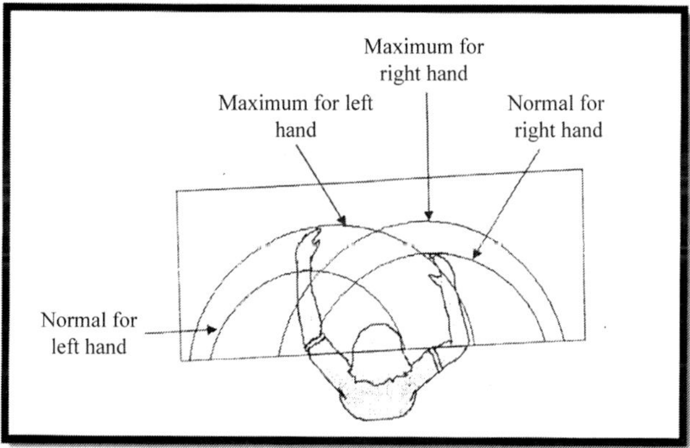

Fig. 2: Normal and Maximum Working Area in Horizontal Plane

Steps to Follows

1. To determine maximum and comfortable work areas at vertical and horizontal plane, first take the height of individual by making him stand besides the wall. The individual should be barefooted, ankle must not touch wall and should be in normal erect position. Usc scale accurately.

2. To find maximum reach for vertical plane, stand facing wall and make an arc on it by keeping the hand away from body as much as you can but do not stretch your body. The point on the arc centre measured from the floor level will give maximum reach at vertical plane.
3. For minimum vertical reach keep your elbow at shoulder height and make an arc. The point at the centre of arc measured from the floor level will give the normal reach at vertical plane.
4. Maximum horizontal reach is measured by sitting on a chair and keep your hands away from your body on the table to the maximum point which can reach without stretching. Now measure that point of your hand to the edge of the table on the side you are sitting.
5. Minimum horizontal reach is measured by keeping the elbows at the side of table and making the arm be flat on the table and measure the point where your hand reaches from the edge of the table on the side you are sitting.
6. In this way, the values for normal and maximum reach at vertical and horizontal plane of each individual are measured and average of all values in each category is taken.

Observation Table: (in cm)

S. No.	Name of Subject	Height (cm)	Vertical plane		Horizontal plane	
			Normal	Maximum	Normal	Maximum
1						
2						
3						
4						
5						
6						
7						
8						
9						
10						
Total Average						

Results: Average of observations (in inches)

1. Average height of the respondents
2. Average minimum range at horizontal plane
3. Average maximum range at horizontal plane

4. Average minimum range at vertical plane
5. Average maximum range at vertical plane

Calculation sheet: ______________________________________

____

____

____

____

____

Conclusion: ______________________________________

____

____

____

____

____

Remarks: ______________________________________

____

____

____

____

____

Siggnature

7

Principles of Doing Work

Objective: To study about principles of doing work.

Introduction

The level of motivation of an individual has a definite relation to the fatigue experienced while doing any work. When a person dislikes a task, motivation to do is low and fatigue sets in faster. Interest in job produces high motivation and prevents fatigue. Besides the way of carrying a task is of great concern if we want to maintain our right posture, in order that we do not have to face any musculoskeletal disorders.

The Principles to Study the Correct Postures

- **Work in Neutral Postures:** Your posture provides a good starting point for evaluating the tasks that you do. The best positions in which to work are those that keep the body "in neutral'. Maintain the "S-curve" of the spine.Your spinal column is shaped more or less like an "S."It is important to maintain the natural S-curve of the back, whether sitting or standing. The most important part of this "S" is in the lower back, which means that it is good to keep a slight "sway back". Working for long periods with your back in a "C-curve" can place strain on your back. Good lumbar support is often helpful to maintain the proper curve in the small of your back. The "Inverted V-curve" creates an even greater strain on your back. Even without lifting a load, bending over like this creates a great deal of pressure on the spine.

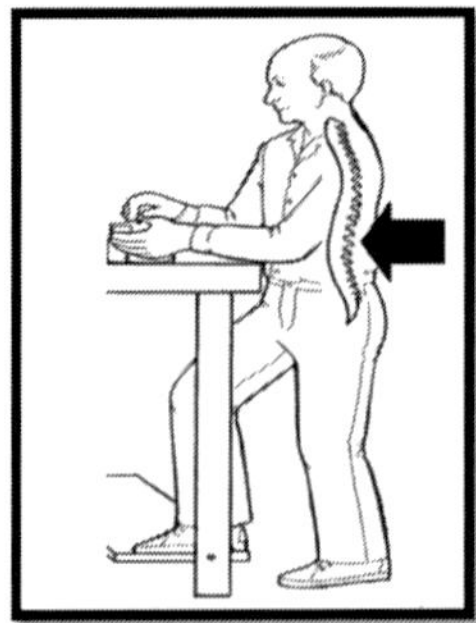

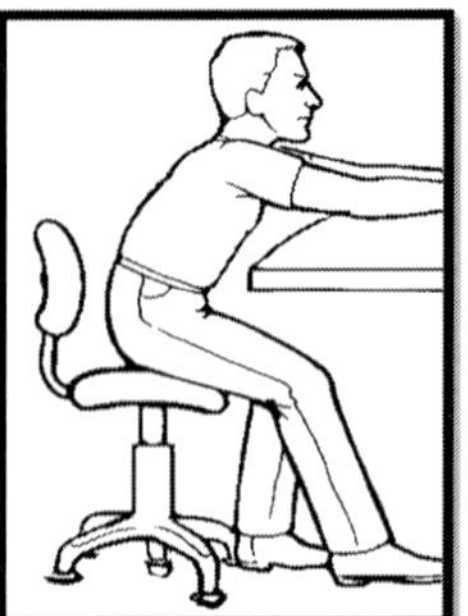

- **Reduce Excessive Force:** Excessive force on your joints can create a potential for fatigue and injury. In practical terms, the action item is for you to identify specific instances of excessive force and think of ways to make improvements. For example, pulling a heavy cart might create excessive force for your back. To make improvements it might help to make sure the floor is in good repair, that the wheels on the cart are sufficiently large, and that there are good grips on the cart..

- **Keep Everything in Easy Reach:** This principle deals with keeping things within easy reach. In many ways, this principle is redundant with posture, but it helps to evaluate a task from this specific perspective. Much of the time, problems with reach are simply matters of rearranging your work area and moving things closer to you. This is not exactly a hard concept to grasp; what is difficult is having the presence of mind to notice and change the location of things that you reach for a lot. Sometimes, the work surface is just too big, causing you to reach across to get something. One option is just to get a smaller surface. Another option is to make a cutout — this way your reaches are cut, but you still have plenty of space for things.

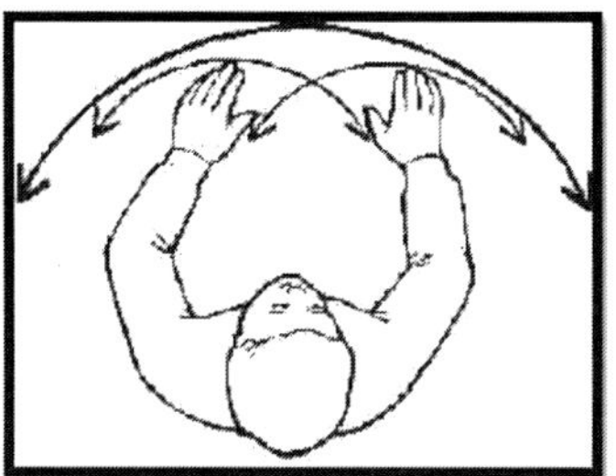
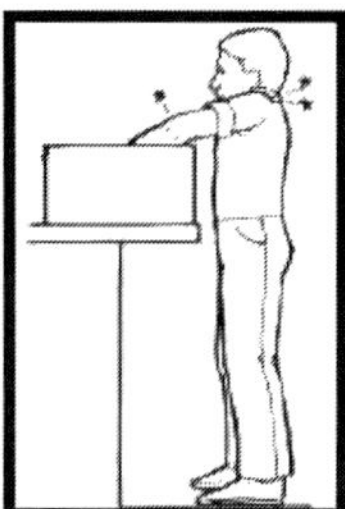

Work at Proper Heights: Working at the right height is also a way to make things easier. Do most work at elbow height? A good rule of thumb is that most work should be done at about elbow height, whether sitting or standing. A real common example is working with a computer keyboard. But, there are many other types of tasks where the rule applies.

- **Exceptions to the rule**: There are exceptions to this rule; however, heavier work is often best done lower than elbow height. Precision work or visually intense work is often best done at heights above the elbow. Or to be a little more complicated, there are ways to make stands and work tables instantaneously adjustable.

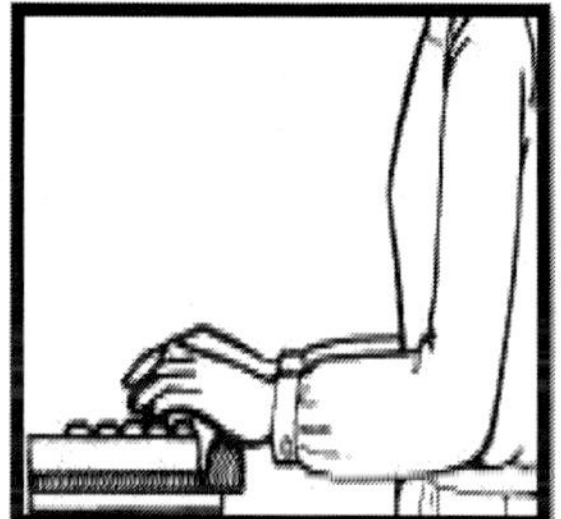
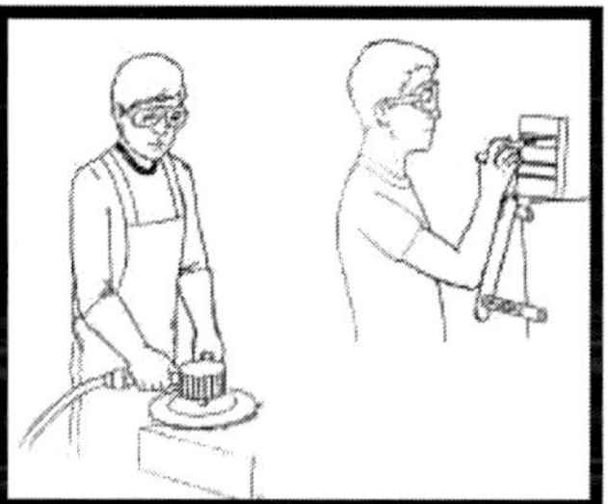
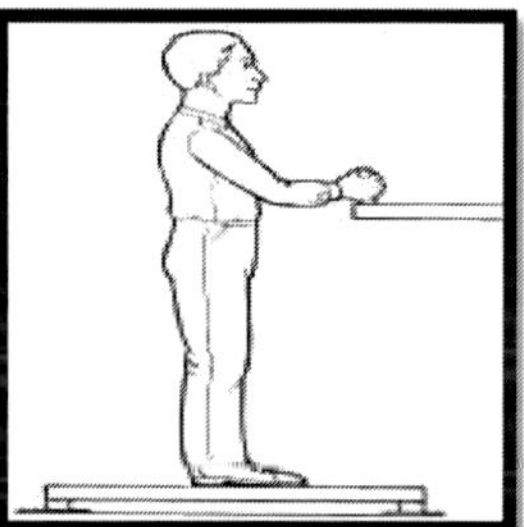

Reduce Excessive Motions: The next principle to think about is the number of motions you make throughout a day, whether with your fingers, your wrists, your arms, or your back. One of the simplest ways to reduce manual repetitions is to use power tools whenever possible. Another approach is to change layouts of equipment to eliminate motions. Objects can be brought closer and tilted, so that you can slide the products in, rather than having to pick them up each time.

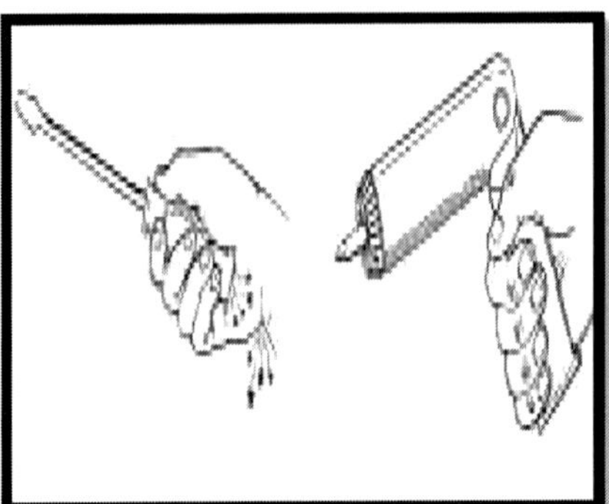

Minimize Fatigue and Static Load: Holding the same position for a period of time is known as static load. It creates fatigue and discomfort and can interfere with work. A good example of static load that everyone has experienced is writer's cramp. You do not need to hold onto a pencil very hard, just for long periods. Your muscles tire after a time and begin to hurt. In the workplace, having to hold parts and tools continually is an example of static load. In this case, using a fixture eliminates the need to hold onto the part. Having to stand for a long time creates a static load on your legs. Simply having a footrest can permit you to reposition your legs and make it easier to stand.

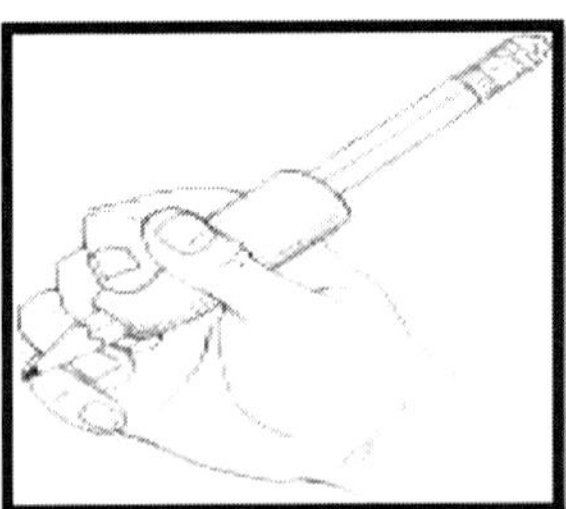

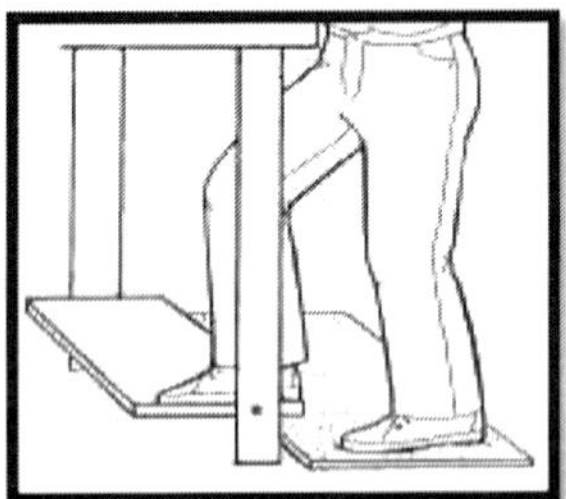

Minimize Pressure Points: Another thing to watch out for is excessive pressure points, sometimes called "contact stress." A good example of this is squeezing hard onto a tool, like a pair of pliers. Adding a cushioned grip and contouring the handles to fit your hand makes this problem better. We've all had to sit on chairs that had cushioning and so understand almost everything we need to know about pressure points. A particularly vulnerable spot is behind your knees, which happens if your chair is too high or when you dangle your legs. Another pressure point that can happen when you sit is between your thigh and the bottom of a table.

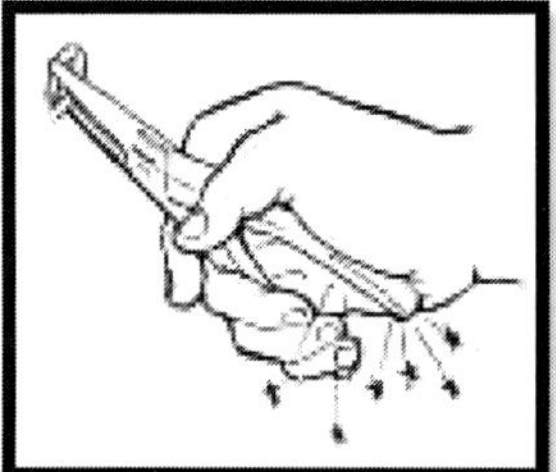

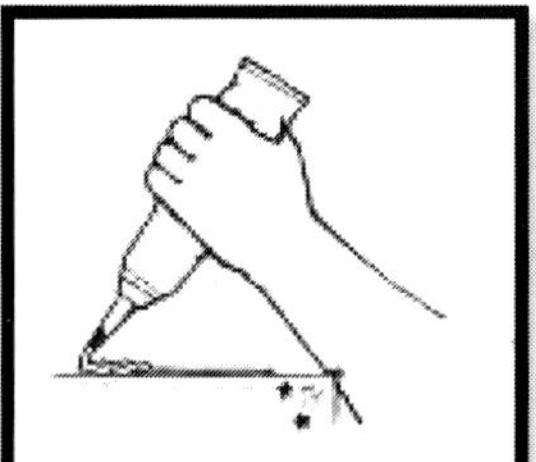

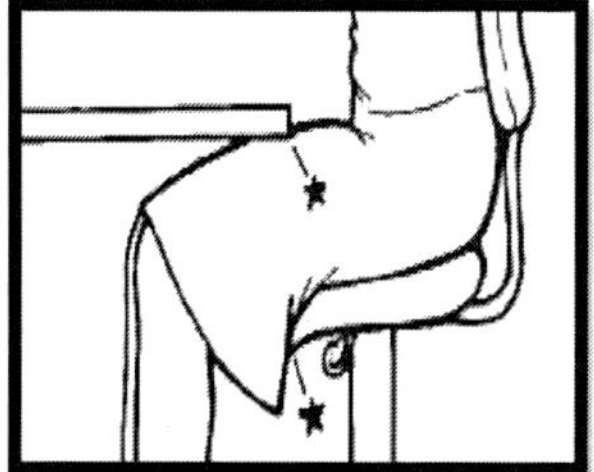

Provide Clearance: Having enough clearance is a concept that is easy to relate to. Work areas need to be set up so that you have sufficient room for your head, your knees, and your feet. You obviously don't want to have to bump into things all the time, or have to work in contorted postures, or reach because there is no space for your knees or feet. Being able to see is another version of this principle. Equipment should be built and tasks should be set up so that nothing blocks your view.

Move, Exercise, and Stretch: To be healthy the human body needs to be exercised and stretched. You should not conclude after reading all the preceding information about reducing repetition, force, and awkward postures, that you're best off just lying around pushing buttons. Muscles need to be loaded and your heart rate needs periodic elevation. Depending upon the type of work you do, different exercises on the job can be helpful. If you have a physically demanding job, you may find it helpful to stretch and warm up before any strenuous activity. If you have a sedentary job, you may want to take a quick "energy break" every so often to do a few stretches. If you sit for long periods, you need to shift postures. Adjust the seat up and down throughout the day.

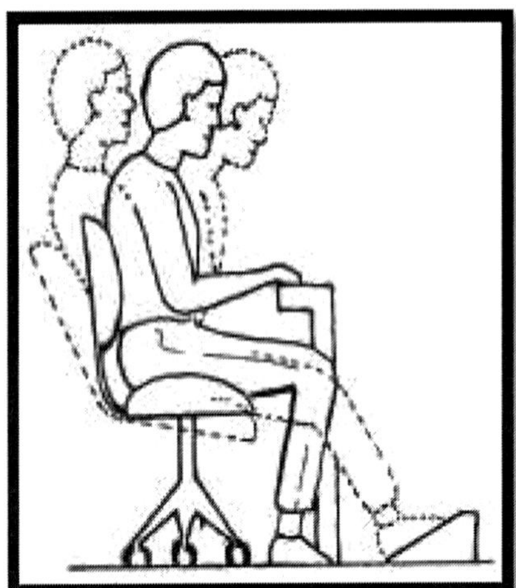

Maintain a Comfortable Environment: This principle is more or less a catch-all that can mean different things depending upon the nature of the types of operations that you do.

- **Lighting and Glare**: One common problem is lighting. In the computerized office, lighting has become a big issue, because the highly polished computer screen reflects every stray bit of light around.

- **Vibration:** Vibration is another common problem that can benefit from evaluation. As an example, vibrating tools can be dampened. Make displays and controls understandable and improving work organization are the most important principles nowadays which we have to take under considerations for reducing any hazards associated with our day to day lives.

Exercise-1

Draw correct posture while working on computers

Conclusion: ______________________________

____

____

____

____

Remarks: ______________________________

____

____

____

____

____

Signature

8

Work Simplification

Objective: To study about work simplification and its techniques.

Work Simplification

Work simplification is the conscious seeking of the simplest, easiest, and quickest method of doing work. It describes the making of daily tasks easier in order to reduce strain or to decrease the amount energy required to complete an activity. Work simplification research consists of making motion and time studies of the work as it is being done, analyzing the work methods, developing the easiest and most effective way to do the task and putting the new method into use. Time and energy are closely related and if these are blended properly the work of the housewife will be much simplified.

Work simplification and motion mindedness are the two important ways of reducing the time and energy expenditure. It simplifies the work methods by analyzing where one can reduce the energy and time usage of the worker. Work simplification means adoption of simplest, easiest and quickest method of doing the work with less consumption of time and energy. Whereas motion mindedness means awareness of the movements involved in doing a particular task and possible ways of reducing the motions.

Work Simplification Techniques

Work simplification is the technique of completing a task by using the least amount of time and energy. As time and energy are closely related, the saving of one leads to that of other. Anyone who is trying to lower time and energy expenditure should learn the value of improving methods of work because the time and energy required to complete a particular work depends on hand and body motions. A convenient method of work must be adopted. There are various techniques for analyzing of work which have proved useful and can be adopted by the home makers in their homes. These techniques can be categorized into two groups:

Formal Techniques	Informal Techniques
• Cycle graph Method • Micro-Motion Film Analysis • Memo motion Study • Stop watch technique	• Process Chart • Operation Chart • Pathway Chart • Multiman chart

Formal Techniques

- **Cycle graph method:** In this method a small electric bulb is attached to the finger, hand or other part of the body of the operator and it is photographed to record the path of the motion. The cycle-graph, a photographic device, is used to study types of motions used in performing tasks. When this is attached to some portion of the body, such as the hand when ironing is being done, it registers the pathway of light projected by a small electric bulb. The resulting record shows whether the movements are smooth and rhythmic or non-rhythmic. This is an effective way to learn how motions may be reduced to improve method of doing task.

- **Memo motion Study:** It was developed 1946 by Marvin E. Mundel at Purdue University, who was first to save film material while planning studies on kitchen work. Micro motion study is the technique of recording and analyzing the timing of basic elements of an operation with the objective of achieving the best method of performing the operation. Short cycle operations require to be studied for microscopic motions e.g., operation of picking up a nut from bin and its fixing consists of three hand motions namely reach for the nut, grasp nut and move hand back to assembly position. Such detailed analysis help to develop the best possible pattern of movements and hence enabling the operator to perform various operations repeatedly with minimum effort and fatigue. Memo motion study is one of the most accurate techniques of work analysis used for work improvement. It makes use of motion pictures of the different activities or movement, so with the help of camera. Very small time upto 0.0005 minute can be measured and recorded by this system.

- **Micro-motion Films Analysis:** Micro-motion film analysis is primarily a research technique and applies best to tasks that can be easily filmed. Motion pictures of tasks done under normal conditions make a permanent record that can be analyzed and charted to show the work of the hands or other parts of the body used in the operation. By means of a timing device, the time of each movement of the worker can be accurately recorded. It is very expensive and not useful to study household tasks. A simultaneous motion chart or simo chart is a right-hand and left hand chart plotted against a time scale and made from micro-motion filming.

- **Stop watch technique:** Stop watch is one of the equipment used for time study. It is employed for measuring the time taken by an operator to complete the work. Stop watch used for time study purpose should be very accurate and preferably be graduated in decimals so that it can recover even up to 0.01 minute.

Informal Techniques

- **Pathway Chart:** The Pathway Chart is a simple device for making a motion and time study in the home. A floor plan drawn to scale and fastened to a drawing board or wallboard, pins and thread are all that are needed to make such a study. Thumb tacks or pins are put in on the floor plan where the worker turns and her pathway is measured from thread wound around the pins as she works. Two charts are used - one for the original and the other for the revised pathway. If measured lengths of string are used, the task of figuring distances traversed is much simplified. Because of tangling when removing any amount of string left over can be more easily measured than the total amount used. If not enough was measured off at the start, additional measured amounts can be used.

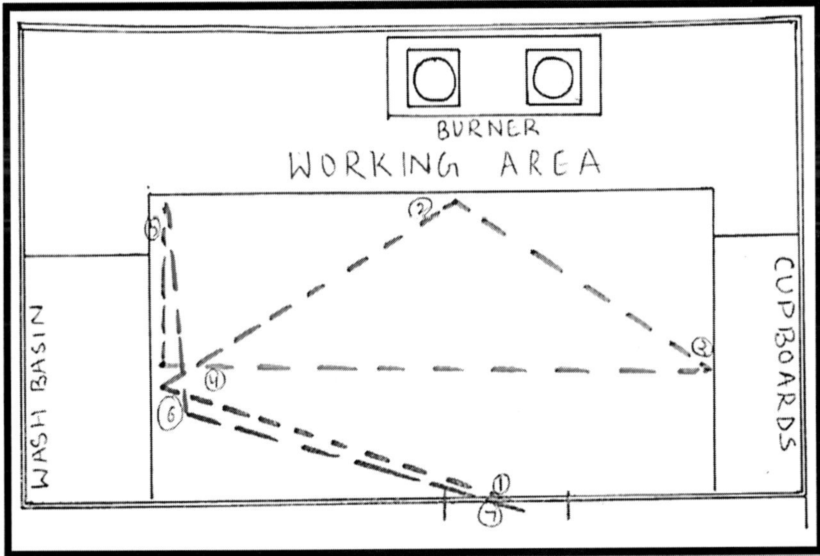

Fig. 1: Pathway Chart

- **Process Chart:** The Process chart is a step-by-step description of a worker performing a given task in its entirety. It is an over-all investigation and differs from an ordinary description of a worker's activities only in that a few symbols are used to clarify the steps immediately into types. Gross, Crandall and Knoll have adapted some of these symbols from industry in the preparation of a process chart.

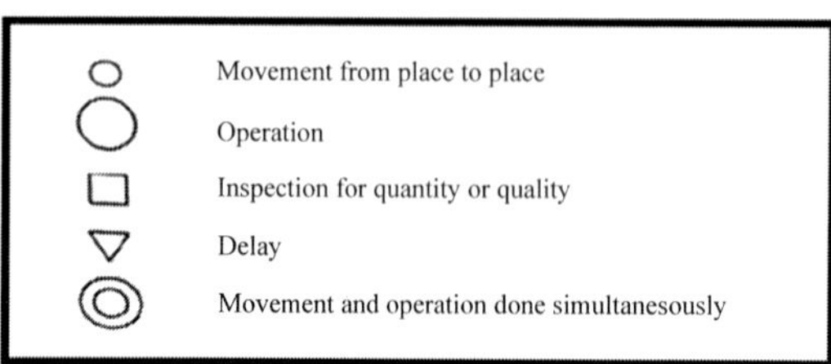

Fig. 2: Symbols used in Process Chart

- The advantage of using the symbols along with the description is that one may quickly count up the number of each type of steps. A composite symbol counts as two activities when summarizing. It requires at least two persons to make a process chart, one to do the task, and at least one other to observe and record. The time is relatively unimportant, as the focus is upon the flow of work. The chart helps to visualize the sequence of an activity.
- **Operation Chart:** The Operation Chart is similar to a process chart, except that it picks up one particular step in a whole process and breaks it down into the work of each hand, shown in parallel columns. The Operation Chart is used in making a more detailed study of some particular part of the process. In this chart, the movements are broken down into the activities of both the right and left hand. The finer analysis show where necessary motions are being made and where delays occur in work. The symbols used are :

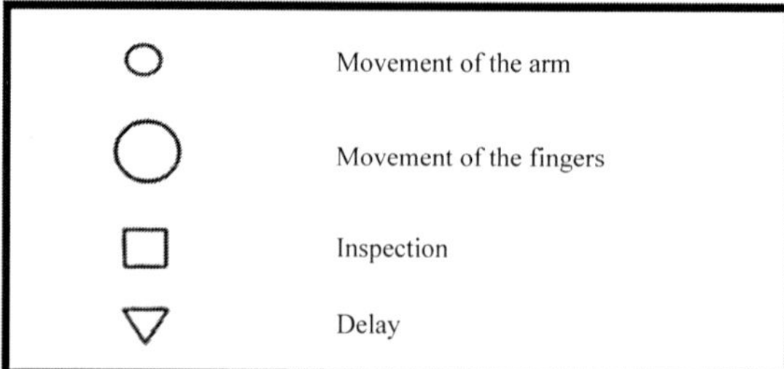

Fig. 3: Symbols used in Operation Chart

- It takes considerably more skill to make an operation chart than a process chart- man analysis. It is practically impossible to make an accurate operation chart of two hands except through film analysis. Each hand can, however be followed by a separate observer and actions of both hands studied, particularly for delays.

- **Multiman chart:** This is used to study activity by two or more people. Delay is an important factor to be watched .e.g. persons making a bed together; each will have certain delays while other is occupied. Thus the labour time for two people making a bed together is greater than the labour time of one person carrying out same task.

Conclusion

In recent years work-simplification techniques have been used by research workers in Home Economics and other fields to improve work methods in homemaking. Motion and time studies have been made of such tasks as food preparation, dishwashing, laundering and ironing, bed making, cleaning and a number of other tasks. Efficient kitchen arrangement, tools, equipment, storage facilities, and correct heights for work surfaces, chairs, and stools have also been studied. These studies suggest many methods of simplifying homemaking tasks, which may be used or adapted by homemakers in their own homes. They show how motion and time studies may be applied to any work problem in the home. Trying out new work patterns and adjusting them to meet new situations add interest to the work. Family members are also kept on the lookout for other ways to make tasks easier, to lessen fatigue, and to make work a pleasure.

Exercise-1

Differentiate the following techniques of work simplification:

Process Chart	Pathway Chart	Operation Chart

Exercise-2

Simplify the tea making activity or work by means of process chart.

Exercise-3

Simplify the Table Setting activity by means of pathway chart.

Exercise-4

Simplify the Ironing activity by the means of an operation chart.

Results: ______________________________

____

____

____

____

____

Remarks: ______________________________

____

____

____

____

____

Signature

9

Kitchen Utensils and Tools

Objective: To study different kitchen utensils and tools used at household level.

Equipments Required: Kitchen tools and utensils used at household level.

Kitchen Utensil and Tools

Home cooking is all about simple, tasty and healthy food brought from the kitchen to the dining table. But oftentimes, preparing the food can be time-consuming and tedious. With the right list of basic kitchen tools, utensils and appliances, we can make cooking more enjoyable and a lot easier. These appliances can also motivate us to think of creative new recipes to delight our loved ones.

- A kitchen utensil is a small hand held tool used for common kitchen tasks such as cutting food items to size, heating food on an open fire or on a stove, baking, grinding, mixing, blending, and measuring; different utensils are made for each task.
- Utensils are built to make your cooking experience as perfect and easy as possible. Most utensils have been invented to make life more convenient.
- Some basic utensils make cooking fun because less time is spent actually preparing the food for cooking. Even a microwave is just piece of kitchen equipment to make heating food easier.
- There are so many utensils that can be used in the kitchen and there is a reason behind that. One of the main reasons for this is the fact that each cooking utensil is fitted to a particular use and if you use the wrong utensil for a certain food, then it might not bring about the same result as the right one.

Types of kitchen utensils and tools

On the basis on convenience and simplification the kitchen utensils and tools can be classified in the following groups:

- Cooking utensils (surface cooking utensils)
- Baking utensils (Oven cooking utensils)
- Measuring tools
- Mixing tools
- Cutting tools
- Preparation tools
- Serving tools
- Storage tools
- Other kitchen tools

Cooking utensils

- **Saucepans and saucepots**: **Saucepans** is a deep cooking pan with a handle and lid for boiling simmering and steaming foods. Whereas a **saucepot** is equipped with two side handles. Pots and Pans are the most essential equipments used in the kitchen. They are made of different materials such as aluminums, steel, iron, brass and copper.

Saucepans

Saucepots

- **Double boiler:** Two pans that fit together so that food can be cooked over boiling water without burning.

Double boller

Pressure saucepans

- **Pressure saucepans:** Pressure pans are usually of the saucepan type with a single handle. Most of them have a domed cover which increases their capacity. They are made of aluminum, cast or stamped, and of stainless steel. Handles and knobs are usually of plastic materials. They vary in size from 2-7 quart capacity, depending on the manufacturer.
- **Skillet fry pan:** A frying pan, frypan, or skillet is a flat-bottomed pan used for frying, searing, and browning foods. It is typically 200 to 300 mm (8 to 12 in) in diameter with relatively low sides that flare outwards. They are made of aluminium or anodized aluminium, cast iron, copper and stainless steel.

Skillet fry pan

Griddle pan

- **Griddle**: A griddle is a cooking device consisting of a broad flat surface. It may be composed of chrome steel, aluminum, or carbon steel.
- **Omelet or Crepe pan** is an uncovered skillet with a narrow bottom and slopping sides. Crepe pans are smaller pans, ranging from 8-12” in diameter that are very shallow with a low rim.

Omelet or Crepe pan

Dutch oven

- **Dutch oven:** The iron kettle known as the Dutch oven. It is a large, heavy pot or kettle with a tight fitting lid used for slow cooking foods on the stovetop or in the oven for long periods of time. The Dutch oven comes in different sizes and has a domed cover which adds greatly to its capacity.

Baking Utensils

- **Cookie/Baking sheets:** These are flat sheets of metal with a low rim on one or more sides. These sheets used for baking cookies, toasting bread. Aluminum, sheet iron, and tinned steel -are the materials commonly used for baking sheets.

Cookie/Baking sheets

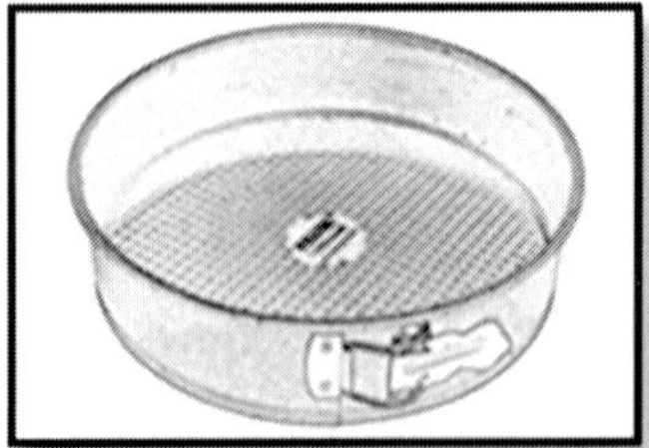

Cake Pans

- **Cake Pans:** Cake pans may be oblong, square, or round; deep or shallow. Various sized pans used for baking identified as width by length in inches. Common ones are: 9”x 13” Oblong, 9”x 9” Square and 9” Round. They are of aluminum, oven glass, tinned steel, iron, or enameled ware.
- **Pizza pans** are large and round and have a narrow rim around the edge. These pans are specialized for making pizzas, similar to a pie pan, cookie sheet, or paella pan, frequently with a perforated bottom.

Pizza pans

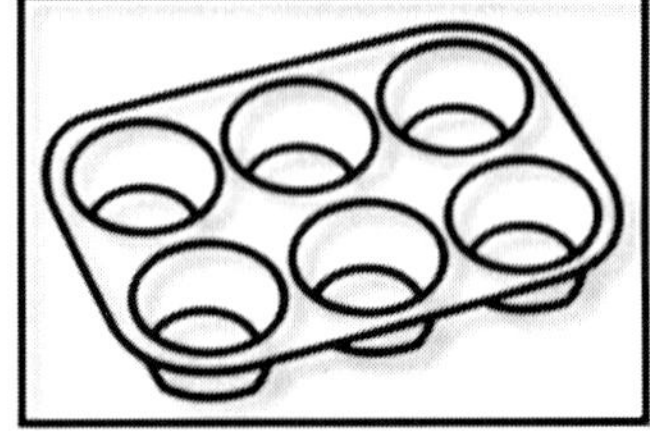

Muffin pans

- **Muffin pans:** These are oblong pans with round depressions for baking muffins and cupcakes.
- **Loaf pan:** A deep, narrow, oblong pan approximately 4 by 9 inches used to bake quick and yeast breads, meatloaf and other items.

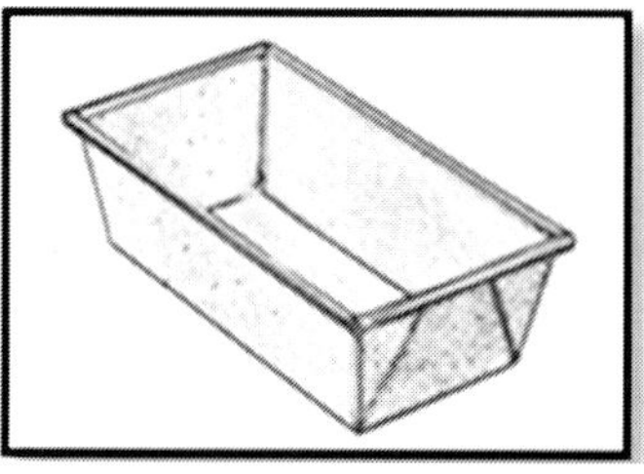
Loaf pan

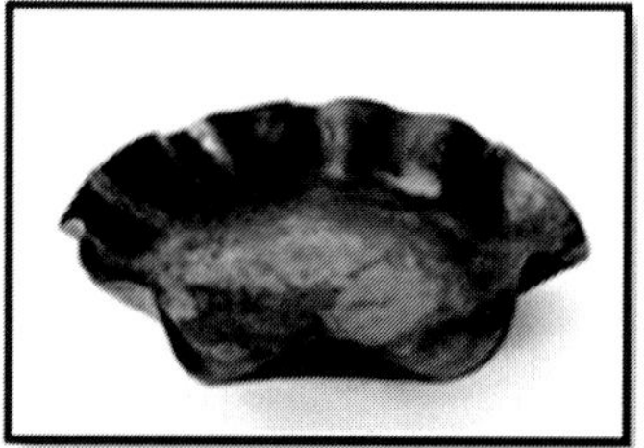
Pie plates

- **Pie plates:** Pie pans are round with sloping sides. Pie pans are of aluminum, tin, enameled steel, and glass, with a flat or juice-catching rim or a scalloped edge to indicate the size of the wedges into which the pie may be cut.
- **Casseroles:** Casseroles are baking dishes with high sides. These are glass, ceramic or metal dishes that may also have a cover to fit.

Casseroles

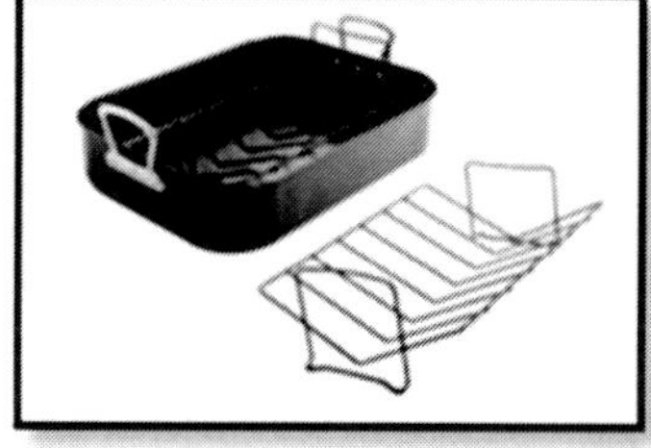
Roasting pans

- **Roasting pans:** Roasters are oval, rectangular, or round, and of enameled ware, aluminum, iron, or steel. Because of its efficiency in absorbing radiant energy in the oven, a dark-surfaced or enameled ware utensil with a large bottom area is preferable to a small, polished pan.
- **Baking parchment:** Parchment paper, or vegetable parchment, is a cellulose-based composite that has been processed to give it additional properties like non-stickiness, grease resistance, and resistance to humidity. It is commonly used in baking as a disposable non-stick, grease resistant surface. It should not be confused with waxed paper, which is paper that has been coated in wax.

Measuring Tools

- **Measuring cup**: A measuring cup or measuring jug is a kitchen utensil used primarily to measure the volume of liquid or bulk solid cooking ingredients such as flour and sugar, especially for volumes from about 50 mL upwards. Measuring cups are of tin, plastic materials, stainless steel, aluminum, or heatproof glass.

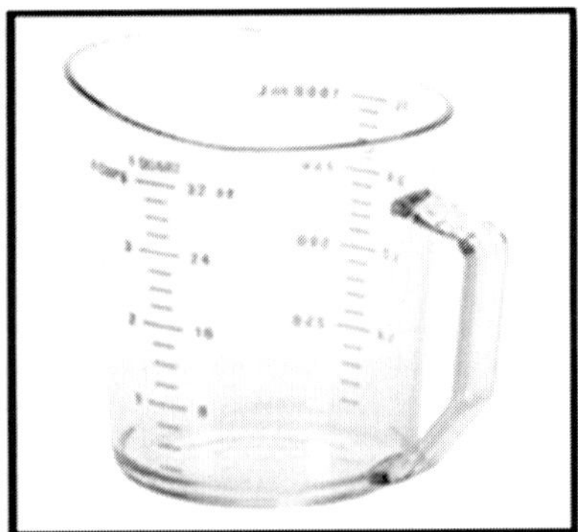

Measuring cup

- **Measuring spoons:** A measuring spoon is a spoon used to measure an amount of an ingredient, either liquid or dry, when cooking. Measuring spoons may be made of plastic, metal, and other materials. Measuring spoons are available in sets, usually between four and six, typically with deciliter (100 ml), tablespoon (15 ml), teaspoon (5 ml) and milliliter measures.

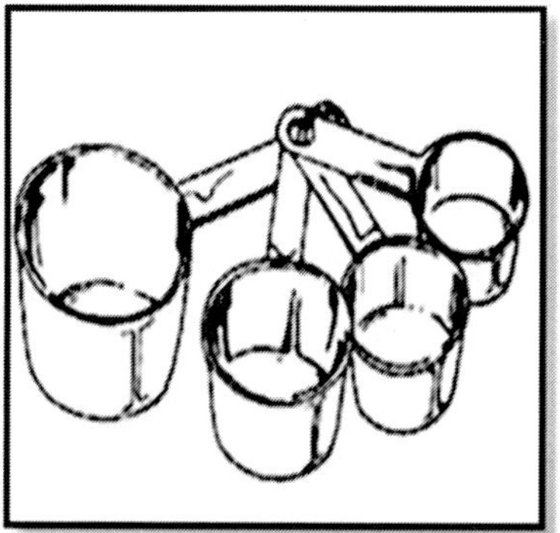
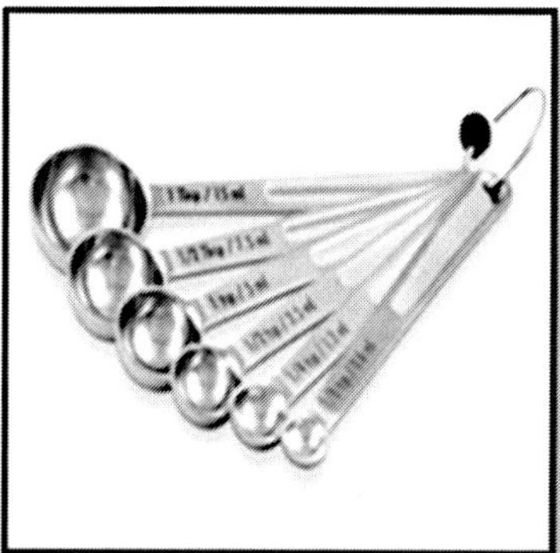

Measuring spoons

- **Kitchen scales:** A kitchen weighing scale is a device to measure weight or mass. These include balance scales, mechanical scales, and digital scales. Digital kitchen scales are used for weighing food in a kitchen during cooking. These are usually light-weight and compact.

Digital kitchen scales

Mixing Tools

- **Spoons**: Spoons for mixing, stirring, and serving are of aluminum, tinned steel, stainless steel, enameledware, iron, inexpensive plated silver or wood.
    - **Wooden spoons** are used for stirring and mixing. Stirring utensil that will not conduct heat as you stir. Probably one of the oldest cooking utensils, which was easily carved from wood with a simple knife.

Wooden spoons

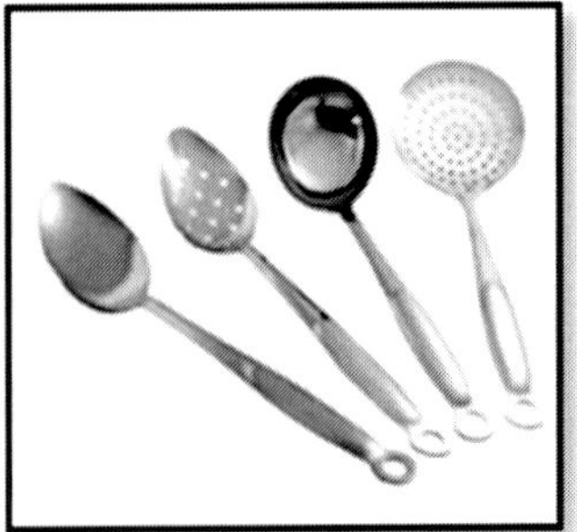

Slotted spoons

- **Slotted spoons** are used to remove pieces of food from liquid. Utensil for lifting solid foods such as vegetables from liquid.
- **Heavy metal spoons** are used to stir thick mixtures

- **Beaters:** The two types of beaters most commonly found on the market are the whisk and rotary.
    - **Whisk**: It is simplest tool for beating and stirring light mixtures. Use it for eggs, souffles, meringues, and sauces.
    - **Rotary beaters** are of steel with four circular or elliptical agitators which revolve in a vertical plane. The agitators are fastened around a support of heavy metal wire, which carries the small pinion wheels. The wire is riveted or welded to the main shaft, to which the large cog wheel and handle are attached. It is used to beat, blend, and incorporate air into food.

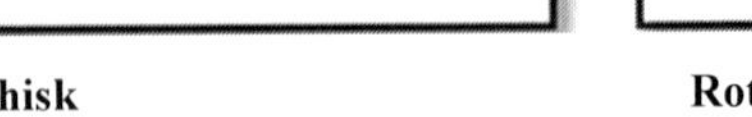

Whisk **Rotary beaters**

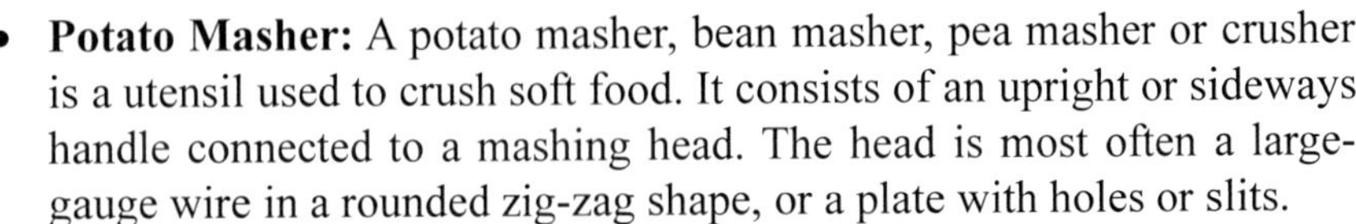

- **Potato Masher:** A potato masher, bean masher, pea masher or crusher is a utensil used to crush soft food. It consists of an upright or sideways handle connected to a mashing head. The head is most often a large-gauge wire in a rounded zig-zag shape, or a plate with holes or slits.

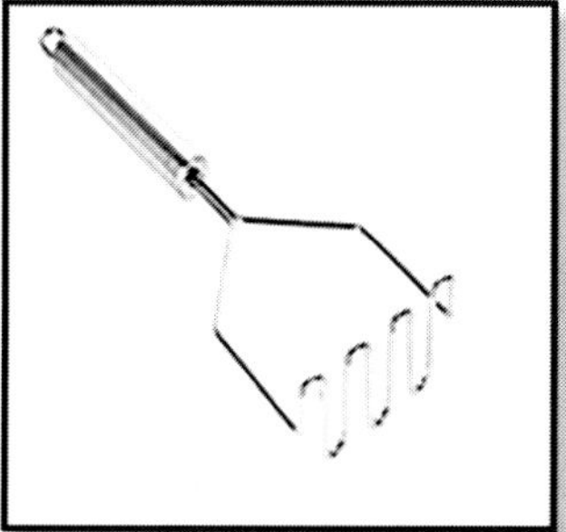

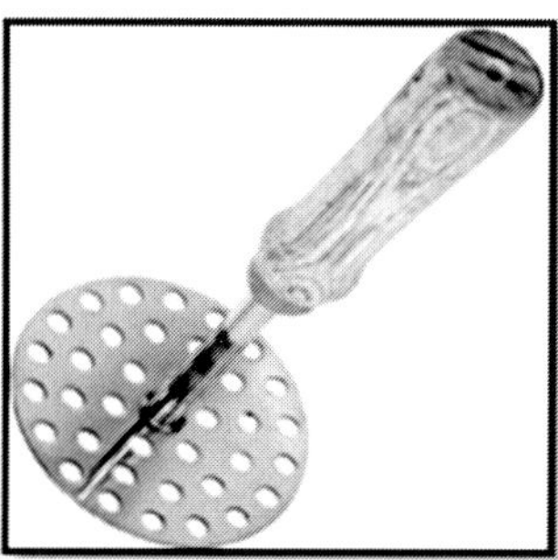

Potato Masher

Cutting Tools

- **Knives:** One of the most basic, but important hand tools for any kitchen is the knife. A knife is a tool with a cutting edge or blade attached to a handle. The type of steel used in a knife blade determines the cutting quality of the blade and how often it must be sharpened. Knives are classified according to the shape and length of the blade, and their names usually indicate the use for which they were made.

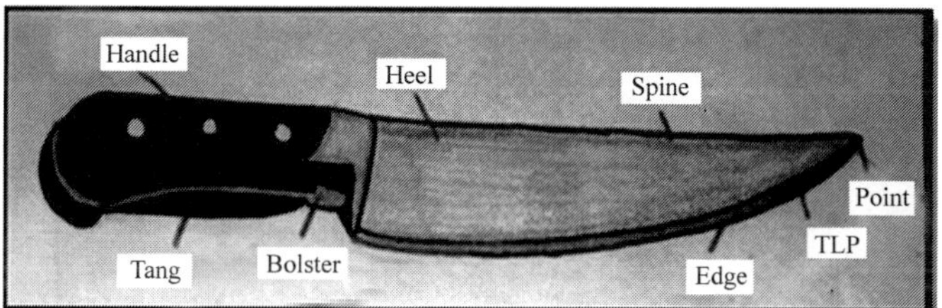

Parts of a Kaife

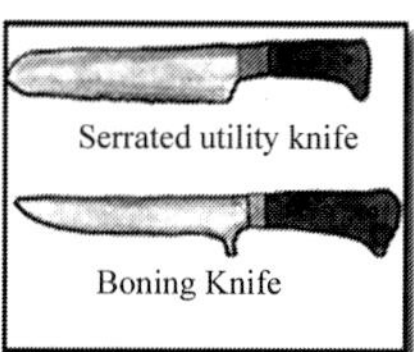

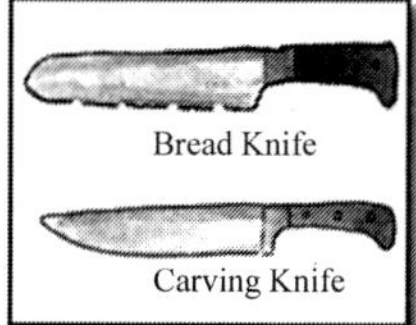

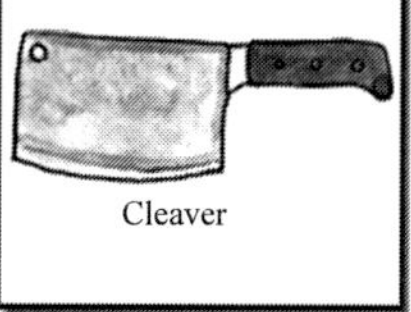

Different types of Knives

- The **French cook knife** also known as chef knife, with its straight edge and firm but tapering point. The handle of the French knife is placed in line with the back of the blade that the hand may not interfere when the knife is used for quantity chopping of nuts and small fruits and vegetables.

- The **paring knife** has a short blade, 2 ½ to 3 inches in length, which gives leverage without undue strain on the finger muscles.
- The **utility knife**, 4 to 6 inches long, may have a straight or curved cutting edge. It has many uses: cutting up large vegetables, trimming meat, slicing cold chicken and cleaning fish-in fact almost any kind of food preparation for which the paring knife is too short.
- **Boning Knife** is super-thin and slightly smaller than a carving knife, giving greater precision, which is perfect for removing bones from meat, poultry and fish.
- Another most useful and efficient knife is the **serrated or scalloped edged bread knife**. The serrated edge has groups of fine ridges, running first in one direction and then in another, which are self sharpening and always cut satisfactorily.
- The **carving knife** has a fairly long sturdy blade, not too pliable, so that it will sever joints and cut hot, yielding meats with ease. The point has a long, rather thin curve, to assist in cutting around the bones.
- The **butcher knife**, with its thick, heavy, curved blade for strength, the grapefruit knife, curved and serrated to cut the sections of the fruit from the rind and the enclosing membranes.
- The **slicing knife** has a long flexible blade, tapering slightly or rounded at the point, and will give the thinnest slices of cold meat, bread or cake.

- **Kitchen Shears:** These are specially-designed scissors for the kitchen. They are not shaped like regular scissors; they are shaped like shears. These used to cut meat, dough, pizza, to snip fresh herbs and trim vegetables. They can also cracks nuts and some shellfish with kitchen shears.
- **Cutting Boards:** A cutting or chopping board is a durable board on which to place material for cutting. The kitchen cutting board is commonly used in preparing food; other types exist for cutting raw materials such as leather or plastic. Kitchen cutting boards are often made of wood or plastic and come in various widths and sizes.

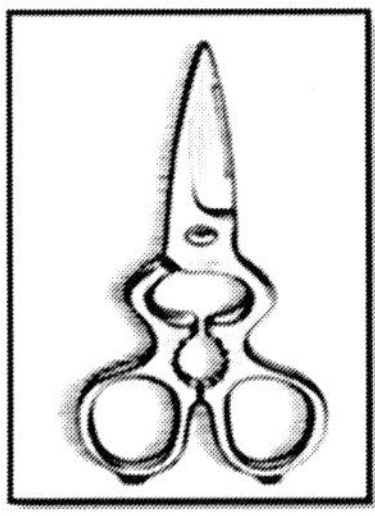
Kitchen Shears

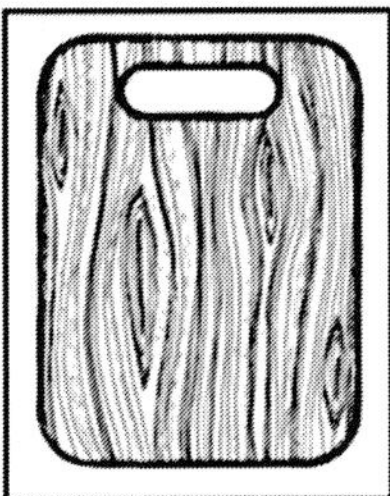
Cutting Boards

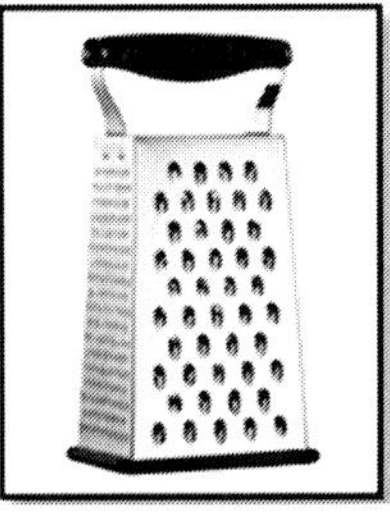
Grater

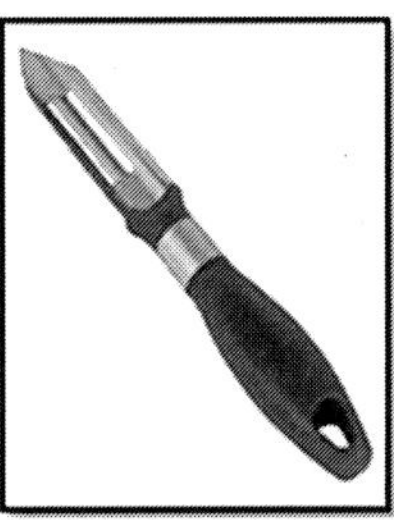
Vegetable Peeler

- **Grater:** A grater may be flat, square, round or cylindrical in shape but the shape is of secondary importance to the type of hole, which is punched or drilled. Graters may be of aluminum or stainless steel, but are usually of tinned steel.
- **Vegetable Peeler:** A peeler is a sharp instrument having a blade and a handle which quickly and efficiently removes the peel from fruits and vegetables without removing any part of the "flesh".

Preparation Tools

- **Spatula:** A spatula is a broad, flat, flexible blade used to mix, spread and lift material. It also refers to a turner which is used to flip over pancakes and meat patties. The blades on these are usually made of metal or plastic, with a wooden or plastic handle to insulate them from heat.
- **Strainers:** Strainers are devices that have holes punched in it or made of crossed wires for removing and draining liquids from smaller amounts of food.

Spatula

Strainers

Colanders

- **Colanders: These** are perforated bowls used to drain fruits, vegetables, and pasta. A bowl with holes used to drain pasta and other liquids from food.
- **Steamer Basket:** A basket which folds and unfolds to conform to the size of a saucepan and keeps the food above boiling water to allow steam to cook the food.

- **Sifter:** Wire mesh and rotating bars for breaking up lumps or loosening flour. Sifters are commonly of tinned steel.

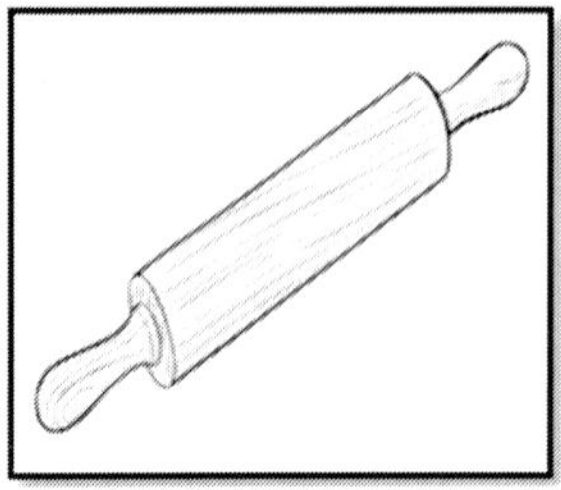

Rolling Pin Steamer Basket Sifter

- **Rolling Pin:** A rolling pin is a cylindrical food preparation utensil used to shape and flatten dough.

Serving utensils

- **Ladle:** A ladle (dipper) is a type of spoon used for soup, stew, or other foods. Ladles are usually made of the same stainless steel alloys as other kitchen utensils; however, they can be made of aluminum, silver, plastics, melamine resin, wood, bamboo or other materials.

- **Pasta fork:** Admittedly, this looks more like an odd-shaped spoon than a fork. However, it's incredibly handy for removing and serving pasta such as spaghetti and linguine that would otherwise go through the holes of a pasta strainer/colander.
- **Pizza cutter**: A curved or circular cutting blade with a handle that serves as a cutting tool to cleanly slice pizza, lasagna, flatbread crusts, brownies, and other similar foods into smaller pieces. This kitchen utensil is typically made of stainless steel and other composite materials. The most common forms of the Pizza Cutters are either the curved Cutter or the wheeled Cutter. Both are designed to slice through the food with only minimal damage to the toppings or top layer.

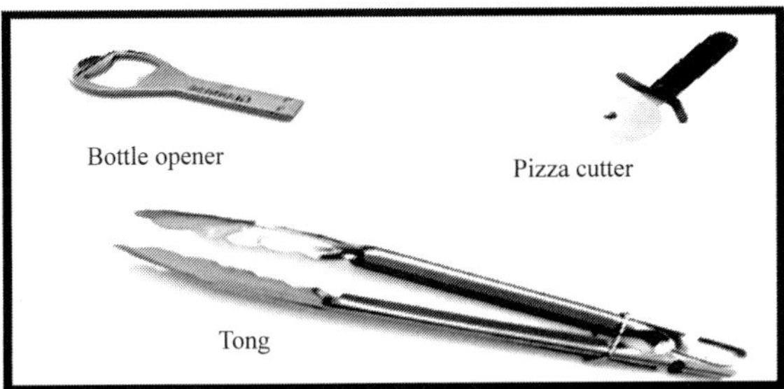

- **Tongs:** Tongs are a type of tool used to grip and lift objects instead of holding them directly with hands.
- **Can Opener:** Bottle openers, handheld crank openers and electric can openers all making opening sealed cans or bottles easier. It is a device that enables the removal of metal bottle caps from glass bottles. More generally, it might be thought to include corkscrews used to remove cork or plastic stoppers from wine bottles. A bottle opener is a specialized lever inserted beneath the pleated metalwork, which uses a point on the bottle cap as a fulcrum on which to pivot.
- **Chopsticks-** Chopsticks used as an eating utensil are made of wood and are long and thin.

Storage Tools

- **Plastic containers**: These are great for keeping food and leftovers fresh in the refrigerator. They can also be used for freezing leftovers in the freezer. Plastic containers are containers made exclusively or partially of plastic. Food storage nowadays relies mainly on plastic food storage containers.

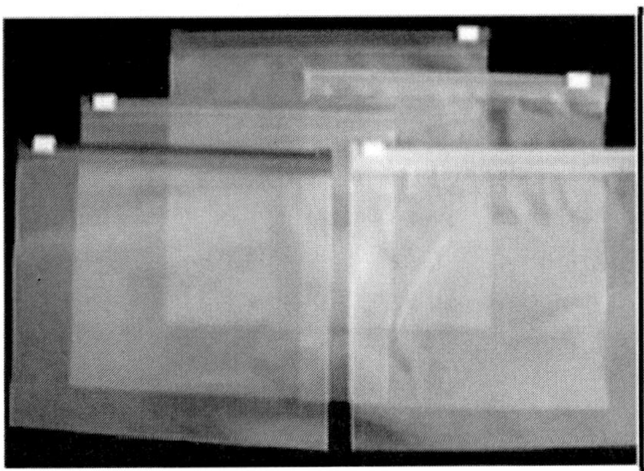

- **Plastic zipper bags**: Great for storing prepared ingredients in the fridge or freezer. These also tend to take up less space than plastic containers but aren't great for liquid. The plastic bags and containers come in different sizes for use with different products. The brand offers sandwich bags, snack bags and other bags for various purposes.
- **Bread bin/box**: Keeping bread in one of these ensures that the bread stays well ventilated, out of direct light, and dry, increasing its shelf life.

Other Kitchen Tools

- **Skimmer:** A skimmer is a flat, sieve-like scoop or spoon used for skimming cooking liquids or lifting ripened cream from milk and for lifting other food items from liquid or oil.

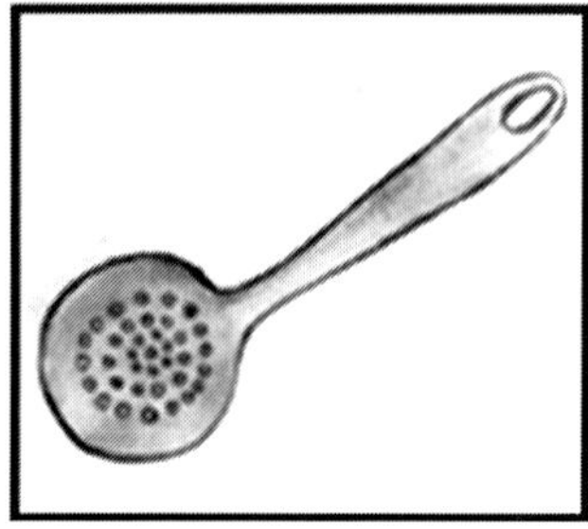

Skimmer

Oven Mitts

- **Oven Mitts and Hot Pads:** Insulated fabric gloves used to protect hands when handling hot items. These will help prevent nasty burns when removing things from the oven.
- **Plastic wrap/Cling film** – These will help keep any food fresh and bug-free during preparation or keeping excess ingredients fresh for longer. Plastic wrap, cling film, Saran wrap, cling wrap, Glad wrap or food wrap is a thin transparent plastic film typically used for sealing or wrapping food items in containers to keep them fresh over a longer period of time. Plastic wrap, typically sold on rolls in boxes with a cutting edge, clings to many smooth surfaces and can thus remain tight over the opening of a container without adhesive.

Clingfilm/warp

Tea towels

Aluminium foil

- **Metal foils**: A thin pliable silvered coloured metal for wrapping and covering foods and for protecting oven roasted joints during cooking. These are great for baking as it reflects heat which can greatly help the cooking process of some food. It is used industrially for a variety of purposes, including packing, insulation and transportation. It's also widely available in grocery stores for household use. At home, people use aluminum foil for food storage, to cover baking surfaces and to wrap foods, such as **meats**, to prevent them from losing moisture while cooking. People may also use aluminum foil to wrap and protect more delicate foods, like vegetables, when grilling them. Lastly, it can be used to line grill trays to keep things tidy and for scrubbing pans or grill grates to remove stubborn stains and residue.
- **Tea towels** – These aren't just for drying up dishes. They can cover food without risk of trapping too much moisture, which is essential in the preparation of some dishes.
- **Pot holders** – these square pieces of fabric make it easy to remove hot pots from the stove and protect your hands in the process. Furthermore, we can place pots on them to prevent the hot surface from damaging our worktops.

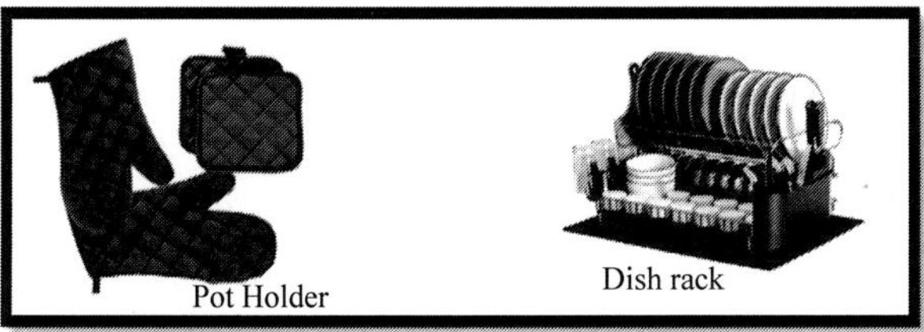
Pot Holder

Dish rack

- **Dish rack-** A dish rack is a kitchen utensil used to hold dishware upright to allow air to circulate to dry or for water to drain.

- **Kitchen/food/meat thermometers:** A variety of thermometers to read the temperature of food or kitchen equipment. These are essential for ensuring good hygiene in our kitchen. Using one of these can ensure that food is cooked properly and all the way through, decreasing the risk of under-cooked food and food poisoning. Very important to use when cooking meat to make sure meat is safe to eat.

Kitchen Thermometers

Exercise-1

Identify the following kitchen tools by their name:

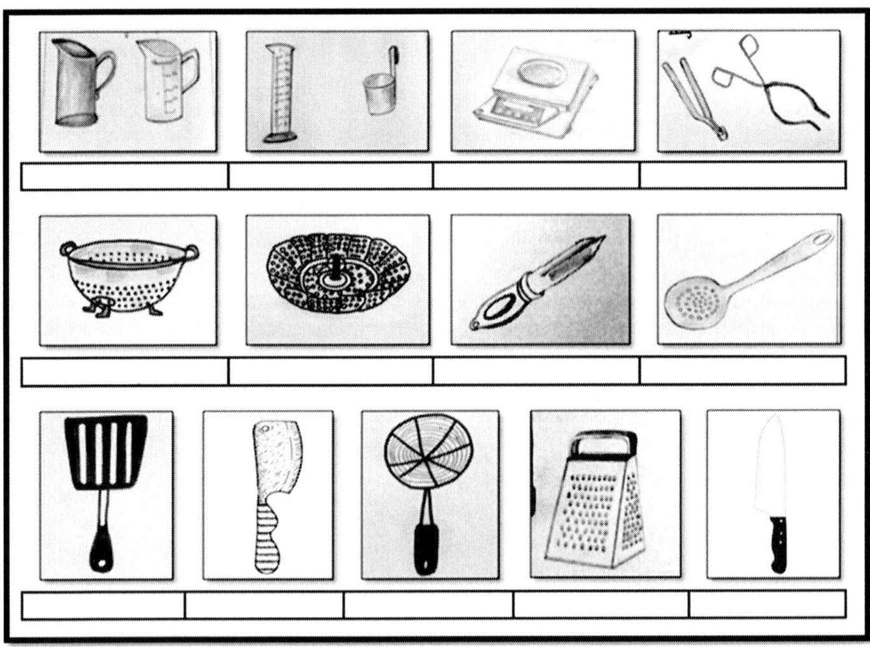

Exercise-2

Identify the following kitchen utensils and tools by their category:

Utensils	Cooking	Baking	Cutting	Mixing	Preparation	Storage	Serving	Miscell-aneous
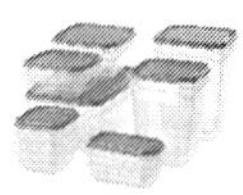								

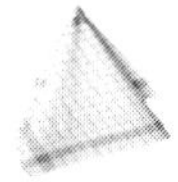								

Conclusion: ______________________________

____
____
____
____
____

Remarks: ______________________________

____
____
____
____
____

Signature

10

Household Equipments and Devices

Objective: To enhance knowledge on household equipments and devices.

Definition of Equipments

Appliances which help or assist in doing household work as cooking, cleaning and food preservation are known as equipments. A equipment also referred to as a domestic appliance or an electric appliance or a household appliance.

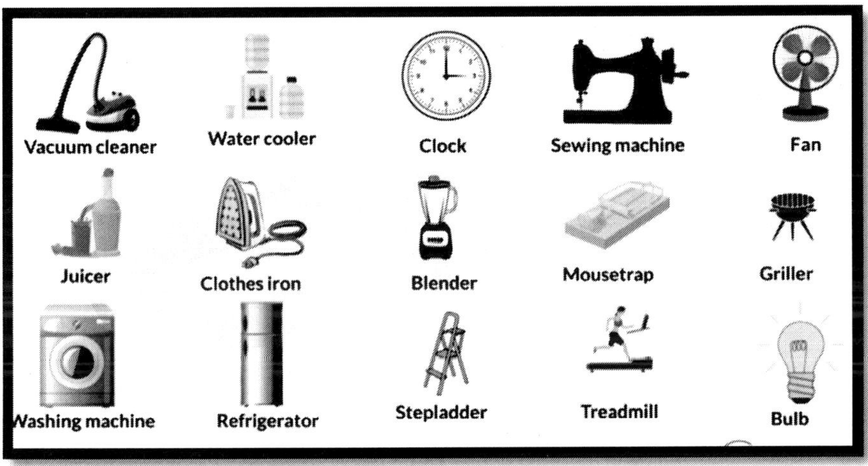

Two Types of Equipments

- **Electrical**: Look around your home and try to identify some equipment which need electricity to work. These are items like toaster, mixer, immersion rod, iron, refrigerator, washing machine, geyser, etc. which cannot work without electricity.
- **Non-electrical**: There is another category of equipment which does not need electricity to run. This category consists of kitchen utensils and tools, sewing machine, cooking stove, solar cooker etc.

Importance of Equipments

- Saves time and energy
- Improves the efficiency of work
- Reduce the drudgery of home makers
- Makes the life comfortable and convenient
- Helps in improving quality of life
- Reduce work of home makers and allows to enjoy leisure time
- Helps in serving variety of food

Some of the Important Appliances/Equipments

Appliance	Description	Picture
Air conditioner	a device that cools and dries the air in one room or an entire home (central air conditioning).	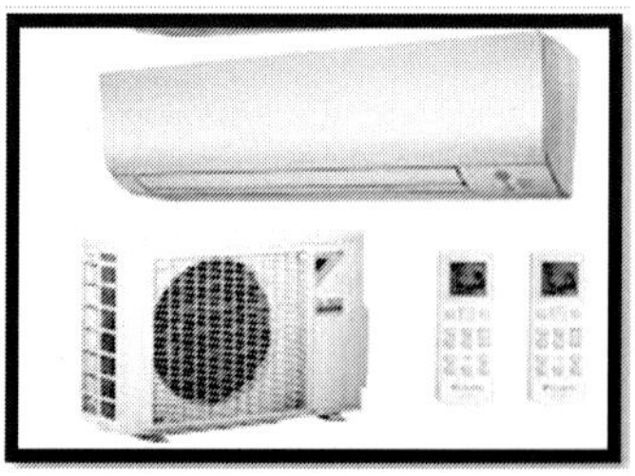
Radiator	heats a room when hot water passes through the device.	
Thermostat/heat pump control panel	a device used to set the temperature of an air conditioner, heater or heat pump.	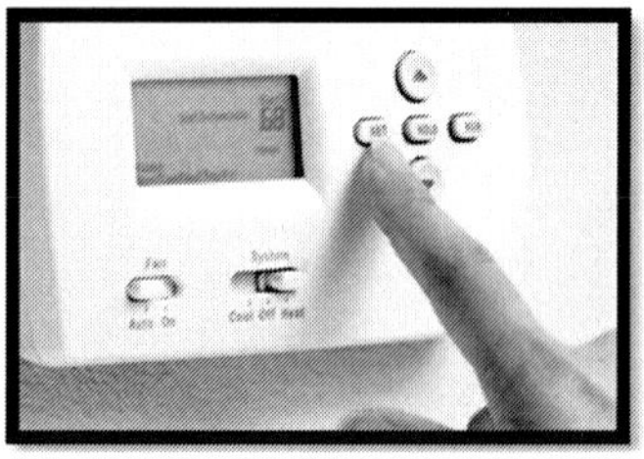

Appliance	Description	Picture
Electric fan	a device with metal or plastic blades that move in a circle and cools and circulate air.	
Space heater	a small electric device that is used to heat a single room or area of a large space.	
Iron	uses heat and/or steam to press out wrinkles in clothing.	
Immersion Rod	made up of an element which becomes hot when electric current passes through it. When we put this immersion rod in water, it heats up the water.	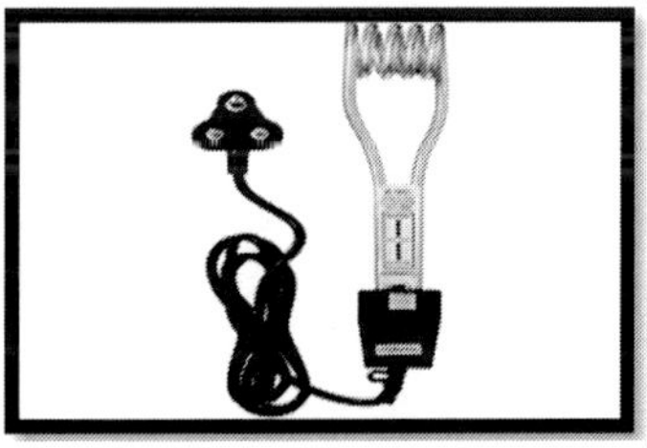
Water Heater/Geyser	a large tank that holds and heats the water used in the house	

Appliance	Description	Picture
Flat screen television	a television that has a flat screen.	
Washing machine	cleans clothes and other materials with water and soap and then spins the water out so that they can dry more quickly.	
Dish washer	a machine that washes and dries the dishes, glasses, silverware, pots and pans.	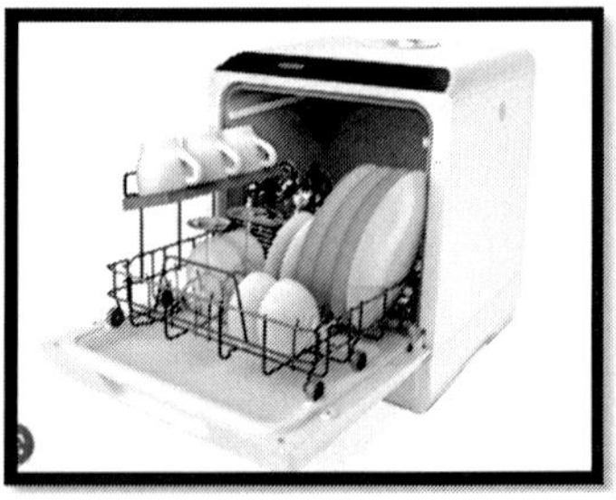
Electric stove or electric range	a stove with an integrated electrical heating device to cook and bake using heat (either powered by electricity or gas).	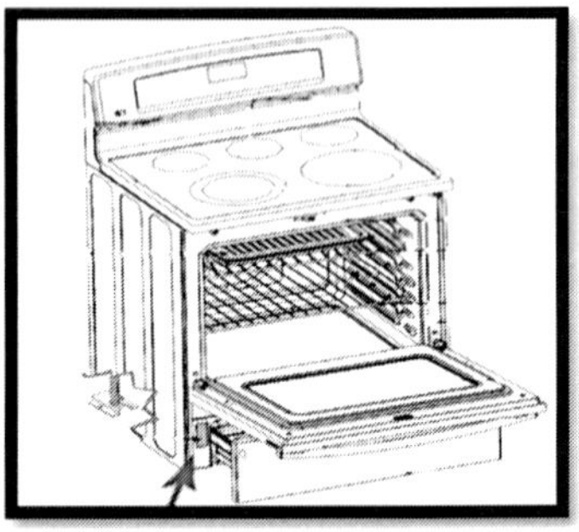
Refrigerator & freezer	an appliance that keeps food preserved with cold air. The refrigerator uses cool air and the freezer uses very cold air to freeze the food.	

Appliance	Description	Picture
Deep fryer	a machine that fries foods in very hot oil .	
Toaster	used to toast bread in order to make it crisp. It has two slots which can hold two or four slices of bread.	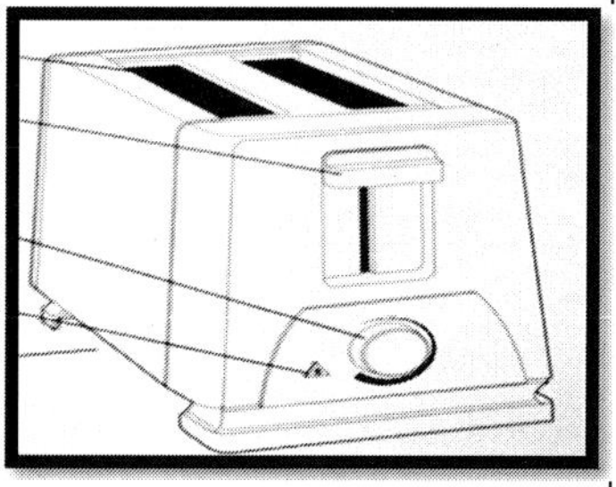
Mixer /Food Processor	an equipment which saves a lot of our time and labour and able to do various tasks like chopping, blending, slicing, whisking, grinding and puréeing, it is like an extra set of hands in the kitchen.	
Microwave Oven	heats and cooks food by exposing it to electromagnetic radiation in the microwave spectrum.	
Electric Kettles/Water Pots	are operated by either a simple on/off switch or feature controls that heat water to a precise temperature to create specialty drinks like tea, coffee, hot cocoa or other warm drinks.	

Appliance	Description	Picture
Electric Rice Cookers	an automated kitchen appliance designed to boil or steam rice.	
Induction Cooktop	uses magnetic currents to directly heat your pots and pans, instead of the cooktop. Cooking with induction makes mealtime easier with quicker heat, precise control, and easy cleanup.	
Coffee Maker	Coffee makers or coffee machines are cooking appliances used to brew coffee.	
Juicer	also known as juice extractor, is a tool used to extract juice from fruits, herbs, leafy greens and other types of vegetables. It crushes, grinds, and/or squeezes the juice out of the pulp.	
Electric Heaters	used to create warmth in cold climates.	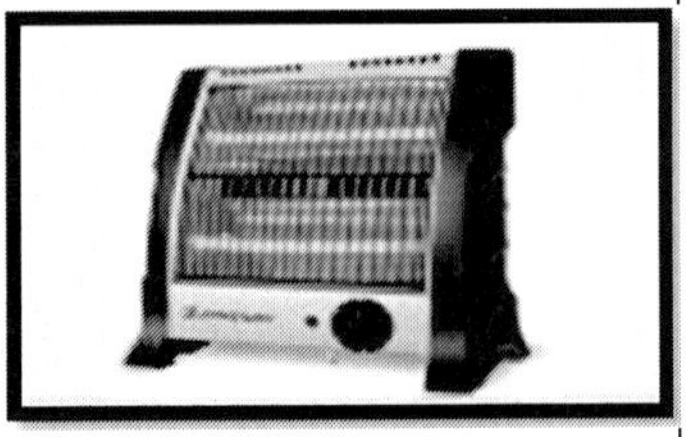

Appliance	Description	Picture
Kitchen Hood	a device containing a mechanical fan that hangs above the stove or cooktop in the kitchen.	
Vacuum Cleaner	a machine used to clean different surfaces like floors, carpets etc. by sucking up crumbs, dirt and other particles.	

Selection of Household Equipments

- **Need-based**: Any equipment you purchase should fulfill your need. Do not purchase an equipment because others have it or it is cheap. For example, instead of buying a food processor with many attachments you can buy a simple mixer - grinder which will fulfill your need. This way you also save money. Time, money and energy saving: When you are buying an item, see that it is useful to you in terms of saving your time and energy as well as your money. For example, a pressure cooker cooks food faster: hence you save time and you also save money because less fuel is consumed. You also save energy because it is simple to use and does not require much supervision. A sharp knife cuts fruits and vegetables easily hence you save your energy. Use of a blunt, cheap knife will not only cut the vegetables poorly, but would also be very frustrating to use.

- **Easy to clean**: The equipments bought should be made of good material which is easy to wash, clean and maintain. For example, iron utensils are difficult to clean whereas stainless steel utensils can be cleaned very easily. A toaster with a removable crumb tray is easy to clean as compared to a toaster fixed with a non-removable tray at the bottom.

- **Safe:** Whatever you buy must be absolutely safe to use at home. Do you know which mark of standardization is used to guarantee the safety of electrical equipment? ISI marked equipment is safe to use because it is properly checked and is of good quality. You must ensure that all electrical equipments that you buy carries this mark. Equipment without this certification may be cheaper but not safe to use. All electrical equipments should have a 3-pin plug and an insulated wire/ cord. Do not select any electrical equipment which has exposed metal parts, as electric current may be conducted through these parts and give you a shock. In the case of non-electrical equipments, see that they do not have sharp edges, loose handles and knobs.

- **Guarantee:** The equipment must also carry a guarantee of service. Guarantee of service means that the manufacturer takes responsibility for the working of that equipment for a specified period of time. The manufacturers of refrigerators may give a guarantee of 5 years, and manufacturers of ceiling fans may give a guarantee of 7 years. This means that for 5 and 7 years your refrigerator and fan should give you trouble free service. If any trouble appears, the manufacturer will repair it free of cost. It is very important for you to ensure that all equipment you buy, specially the costly equipment, should carry a guarantee of service. Ask the shopkeepers about those parts of the equipment which carry guarantee so that you are not cheated later. At the time of buying, see that the guarantee card is duly filled in and signed and stamped with the seal of the shopkeeper from whom you are buying the equipment.
- **Cost:** Do comparative shopping and find the equipment available at better price. In addition to investment cost adds operation and maintenance cost. While some equipments which are simple do not cost too much, there are some which have more complicated parts or attachments and are quite costly. They are expensive because they perform more functions. For example, a refrigerator is very costly whereas a "grameen sheetal" is much cheaper. But when used, a "grameen sheetal" will not be able to do all that a refrigerator can. You can get ice, keep food and water cool, make icecreams, preserve food, etc., in a refrigerator but not in a grameen sheetal. A grameen sheetal will last for a few years only whereas the refrigerator will last for many years. Hence, after seeing the advantages, provided one can afford it, one would be quite right in selecting the costlier refrigerator. This is not true in all cases. For example, it would not be advisable for a housewife to buy a complicated and expensive model of a sewing machine when she needs to do only simple repair work at home.
- **After-sales service**: When equipment is constantly used, it is bound to undergo wear and tear in due course of time. If you need any service, repair or replacement of parts of an equipment, then it should be conveniently available at an affordable price. Therefore it is necessary

to make sure that after sales service is available at the local market place. Equipment with a good after-sales service is always the right choice!

General Safety Guidelines

1. Use electrical appliances with safe 3-pin plugs (except those not supplied through socket outlets).
2. Arrange registered electrical workers to install electrical appliances not supplied through socket outlets. Check electrical appliances, including their plugs and flexible cords, for any damage before use. Replace plugs by qualified persons only.
3. Buy appliances with guarantee card stamped by the dealer.
4. Keep all appliances away from heat, water and moisture.
5. Do not use any electrical appliance if its safety or proper operation is in doubt, and have it inspected by qualified persons.
6. Switch on an electrical appliance only after plugging it into a socket outlet. Otherwise, sparks occurring at the plug may cause fire hazard.
7. Always disconnect appliance after use.
8. Allow sufficient space for ventilation to prevent electrical appliances from overheating.
9. Keep children away from operating electrical appliances.
10. Thoroughly clean all household equipment after every use with due precautions.
11. Wipe the outside surface of an equipment to prevent deposits of dust and grease.

Exercise-1

Draw two electrical and two non-electrical appliances used at household level.

Electrical	Non-Electrical

Conclusion: ____________________

Remarks: ____________________

Signature

11

Cutlery (Spoons) and Its Uses

Objective: To identify the cutlery (spoons) and its uses.

Introduction

Today, we have spoons made of different sizes and materials such as metal, wood, plastic, ceramic or a combination of two or more materials. The most popular material is the stainless steel. Spoons are used primarily for eating liquid or semi-liquid foods, such as soup, stew or ice cream, and very small or powdery solid items which cannot be easily lifted with a fork, such as rice, sugar, cereals and green peas.

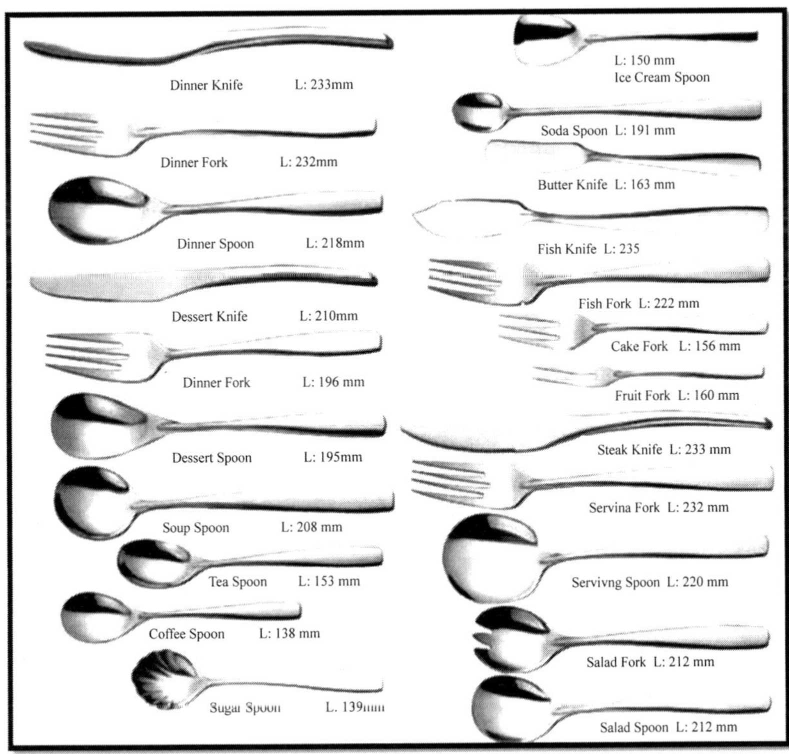

1. Table Spoon 8.0” - 8.5”

Larger than a teaspoon, a tablespoon is both a unit of measurement and a utensil that is used regularly during meals. We can buy tablespoons that are shaped and designed specifically for measuring ingredients; tablespoons are also utensils that are used for eating. They are ideal for adults to use due to their larger size and make it very easy to eat our meal. While some people think that they are perfect for eating soup, there is actually a dedicated soup spoon available that has many benefits over this shape of a spoon.

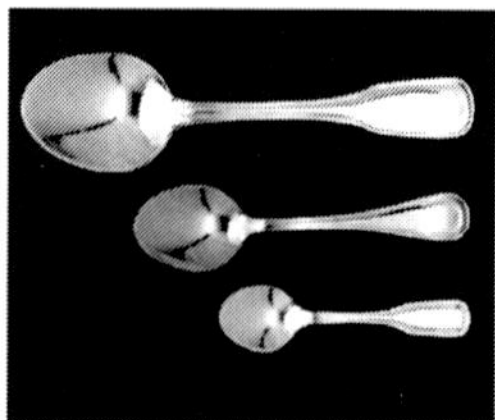
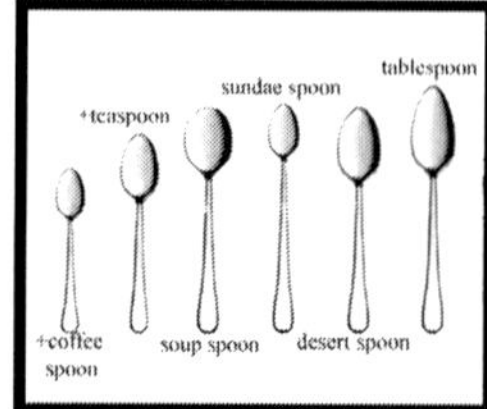

2. Teaspoon

Teaspoons are smaller than tablespoons and almost look like a baby spoon. Teaspoons have a similar size to the measuring teaspoons people use for measuring ingredients, cooking, or baking but they are actually smaller.

3. Soup Spoon (7.0” - 7.5”)

It is a bit deeper in shape making it easy and comfortable for a person to scoop the soup without making a mess. There are also two types of soup spoons, differentiating from their size and styles. The two main variations are the British and Chinese soup spoons. It’s easy to distinguish both as British spoons are small and the bowl is circular in shape. On the other hand, Chinese soup spoons vary in size from small to extremely large ones. They have an odd look handle that is thick and extends from the bowl. Most Chinese soup spoons are made of ceramic but you can also find some that are made from different materials.

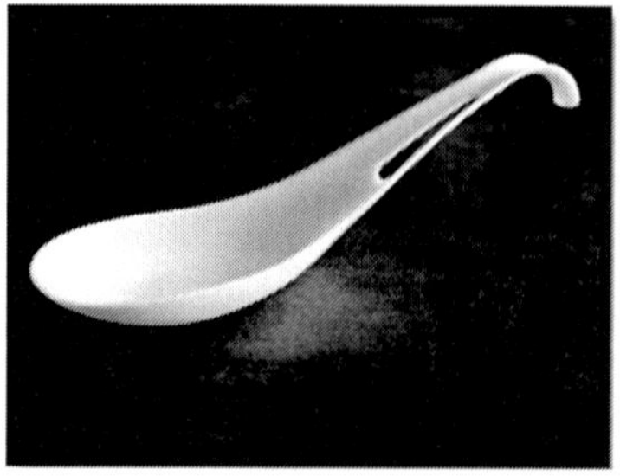

4. Sugar Spoon (5.5" - 6.0")

Sugar spoons have a long, narrow cup with slightly raised edges. This special cup helps for pouring the sugar into small-sized cups without spilling it. Sugar spoons are sometimes called "**sugar shovels**" because of their rectangular shape and deep bowl. What is the purpose of a sugar spoon? A sugar spoon is specifically designed and designated for **scooping table sugar**. A sugar spoon is a piece of formal silverware or flatware that is typically a part of a tea or coffee service.

5. Coffee Spoon (4.5" - 5.0")

This is a small spoon designed for stirring Italian-style "Espresso" coffee prepared with a coffee mocha machine, the Neapolitan coffee maker or coffee-shop machine. Its small size serves for small coffee cups and glasses.

6. Cream Spoon (4.5" - 5.0")

Cream spoon has a rounded cup that is perfect for serving sauces or cream onto other food.

7. Dessert Spoon (8.0" - 8.5")

This spoon is specifically designed for having desserts. It has got an oval bowl and is almost double the size of a teaspoon.

8. Ice cream Spoon (6.0" - 6.5")

These spoons have a wide-cup that is typically square in shape for picking up just the right amount of ice cream from the bowl.

9. Olive Spoon (6.0" - 6.5")

The olive spoon is used to serve olives and other appetisers preserved in liquids. The spoon's cup has holes for the liquid to drain through, making it easier to serve the food.

10. Rice Spoon (10.0" - 11.0")

This is a large spoon designed for serving rice dishes and other side dishes at the table. The cup of this spoon is very large and can hold a larger amount of food.

11. Sundae Spoon / Parfait Spoon (9.0" - 10.0")

Sundae spoons are characterized by a very long handle. These spoons are designed for stirring long drinks, cocktails and any drink served in a tall glass. The tapered shape of the spoon allows it to reach the bottom of any glass.

12. Fruit Spoon

Fruit spoons are smaller in size compared to tablespoons. These spoons have pointed tips and they also have an elongated bowl that helps them cut the fruits easily.

13. Long Drink Spoon

These drink spoons as the name suggest have a very long handle. These spoons are designed to stir hot and cold drinks from long glasses or big mugs.

14. Salad Spoon

The classic salad server set usually consists of one spoon-like and one fork-like implement, but these days there are a dizzying number of options to choose from, including short, chubby salad 'hands'. But for some, a pair of tongs is the way to go.

Exercise-1

Draw different types of spoons and mention their uses:

Conclusion: ______________________________

____

____

____

____

____

Remarks: ______________________________

____

____

____

____

____

Signature

12

Pressure Cooker

Objective: To study the parts, functioning and performance of pressure cooker.

Introduction

Pressure cookers are pans made of Aluminum or stainless steel. They are available in different sizes. The pressure cooker was invented in the seventeenth century by the physicist Denis Papin, and works by expelling air from the vessel, and trapping the steam produced from the boiling liquid inside. This raises the internal pressures and permits higher cooking temperatures. This, together with high thermal heat transfer from the steam, cooks food far more quickly, often cooking in between half and a quarter the time for conventional boiling. After cooking, the steam pressure is lowered back to ambient atmospheric pressure, so that the vessel can be opened safely. Almost any food that can be cooked in steam or water-based liquids can be cooked in a pressure cooker.

Parts and Functioning

A pressure cooker is essentially a vessel of stainless steel or aluminum alloy. It has a lid fitted with a safety valve at the top to prevent the internal pressure from becoming greater than the vessel can bear. A rubber ring is fixed at the junction of the vessel, which prevents the steam from leaking out. A heatproof handle is attached to the pressure cooker for handling the vessel safely. A safety device on the cover of the pressure cooker is so designed that it blows out if the pressure becomes too high. A rubber gasket on the lid releases excess pressure if the cooker goes dry.

A pressure cooker is based on the principle that with the increase in pressure, boiling point of water increases. It works by capturing the steam which increases the pressure in the vessel. Due to increase in pressure the boiling point of water rises up. When the water inside the pressure cooker boils, the inside pressure is designed to build up to 1kg/sq.cm and at this pressure, the corresponding temperature is 120°C as against 100°C , which is the boiling point of water in conservative cooking. Thus, it increases food cooking speed upto four times. This helps in saving fuel.

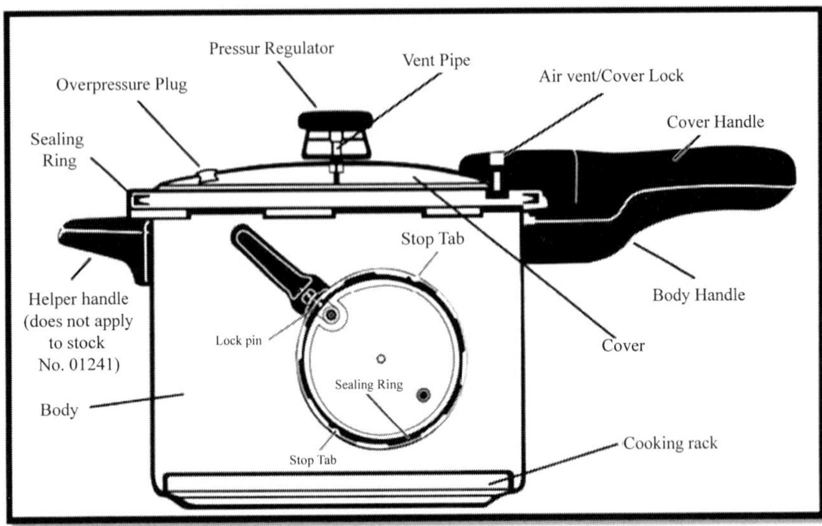

Types of Pressure Cookers

There are two types of traditional pressure cookers: Stovetop and electric pressure cookers:

- **Stovetop pressure cookers** made of stainless steel and have a thick bottom that helps disperse the heat evenly so that the pressure rises appropriately. Cook food faster and are sturdier than electric cookers but require a bit more attention while cooking, because the heat source for these pots is external, it's necessary to manually adjust and monitor the heat to maintain pressure levels during cooking.

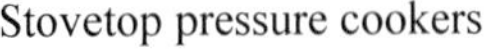

Stovetop pressure cookers

Electric pressure cookers

- **Electric pressure cookers** are slower and less durable but easier to operate, allowing users to automatically adjust the heat and making the procedure nearly hands-free. They also switch to a keep warm setting after the food is done. The control pads are simple to use and typically fail-safe. Many electric pressure cookers have useful features like as programmable start times and warming settings, as well as the ability to perform various cooking tasks such as slow cooking and steaming rice.

Safety Precautions While Using

- Read all Instructions before use.
- The hole from where the steam comes out must not be blocked; otherwise the pressure cooker can burst.
- Do not lift the weight to let the steam out of the pressure cooker as it can burn your hand.
- The gasket must be regularly checked, and promptly replaced if it becomes loose.
- Before putting the Weight Valve device on the cooker it should be ensured that the steam starts to come.
- Do not touch hot surfaces. Use handles or knobs.
- Close supervision is necessary when the pressure cooker is used near children.

Advantages and Disadvantages

- **Advantages of using Pressure Cooker:** The advantages of using a pressure cooker are that they reduce the cooking time by almost two third, and also there is less loss of nutrients.
- Disadvantages of using Pressure Cookers: Disadvantages of using a pressure cooker include that it does not allow you to cook different items at same time. Also, time taken by a pressure cooker to cook at higher altitudes is different.

Exercise-1

Write down any five key features of the following cookers:

Stovetop Pressure Cooker	Electric Pressure Cooker

EXERCISE-2

Prepare rice in stovetop and electric pressure cooker to test the performance of both cookers and evaluate the prepared product on the 9-point hedonic scale:

1-Like Extremely

2-Like Very Much

3-Like Moderately

4-Like Slightly

5-Neither Like nor Dislike

6-Dislike Slightly

7-Dislike Moderately

8-Dislike Very Much

9-Dislike Extremely

Testing of Prepared Food Products on Hedonic Scale

S.N.	Stovetop Pressure Cooker					Electric Pressure Cooker				
	Appearance	Texture	Aroma	Color	Taste	Appearance	Texture	Aroma	Color	Taste

Results

- **Best Appearance:**
- **Best Texture:**
- **Best Aroma:**
- **Best Color:**
- **Best Taste:**

Conclusion: ____
____
____
____
____
____
____

Remarks: ____
____
____
____
____
____

Signature

13

Pots and Pans

Objective: Testing of different types of pots and pans.

Introduction

Cookware is food preparation equipment, such as cooking pots and pans, used in kitchens. There are several different types of pots and pans- **nonstick, ceramic, stainless steel, cast-iron and more**. If you're only going to buy one, we suggest a heavy-duty, try-ply or multi-clad stainless steel pan. They're expensive, but they'll last a lifetime with proper use.

Materials Used

These should be made of thick high-quality stainless steel and have a heavy bottom that ensures even heating. This is important especially for making delicate egg-based custards but also comes in handy for everyday cooking tasks like making oatmeal and rice

- **Stainless Steel:** Long-lasting, classic, uncoated stainless steel is a good choice for browning and braising. Stainless steel pans and surfaces are the best for browning ingredients and since they're usually uncoated, unlike nonstick varieties, they are more durable and resistant to slip-ups in the kitchen.

- Stainless steel pans look smooth, but the cooking surface actually has tiny pores. When you heat the pan, the steel expands and the pores shrink. The shrinking pores grip onto the food, causing it to stick. To prevent sticking, preheat the pan to medium, then add ample oil, then add the food.
- **Nonstick:** Durable nonstick coatings effortlessly release even delicate foods, including eggs and pancakes. Non-stick pans typically last between one and five years. Non-stick pans coated with PTFE (Teflon) generally last at least three years, and ceramic-coated non-stick pans last an average of two years. Pans with multi-layer PTFE coating or reinforced non-stick coating last the longest.
- **Enameled Cast Iron**

Different Types of Pots and Pans

- **Casserole pan:** A casserole pan is similar to a Dutch oven but with shorter sides. The most versatile models are made with cast iron and contain a lid, allowing you to use them on the stovetop or in the oven. They maintain steady heat and are great for low-and-slow cooking, like stews or oven-baked casseroles.
- **Crepe pan:** These pans are very similar to egg pans and frying pans, but they have flat edges instead of fluted sides. It's so much easier to get a spatula under a pancake or crepe when the pan's lip doesn't get in the way! They're available with and without nonstick coating, but we prefer the latter because they're easier to use for beginners.
- **Dutch oven:** If you're looking for one pan to make soups, sauces, stocks, braises and more, the Dutch oven is it. You can even deep-fry food and bake bread, chicken and pasta in these tall-sided pans. They're slow to heat up but retain heat well, thanks to their cast-iron construction. Some models contain porcelain-enameled interiors for added nonstick and ease of cleaning.
- **Egg pan:** An egg pan is essentially a small frying pan, but they always contain nonstick coating. There are a few different nonstick materials these days, so check out our nonstick pan buying guide to find the best one for your kitchen. We like 8- or 10-inch egg pans because they're small enough to fry a single egg but large enough to make a killer omelet.
- **Frying pan:** A frying pan is made with a long handle and wide, sloped edges. They're a kitchen powerhouse for sauteing vegetables, building pasta sauces or pan-frying food. There are several different types of

frying pans — nonstick, ceramic, stainless steel, cast-iron and more. If you're only going to buy one, we suggest a heavy-duty, try-ply or multi-clad stainless steel pan. They're expensive, but they'll last a lifetime with proper use.

- **Griddle:** A griddle is a large, flat cooktop without sides or edges. It's great for breakfast items like eggs or pancakes, but you can also use it over high temperatures for searing steaks or burgers. Most griddles are large enough to straddle two burners on your cooktop, and our favorites are two-sided to include a grill pan. Depending on your range, you may not be able to use a cast-iron griddle, but luckily there are several electric griddles available today.
- **Paella pan:** These pans are specifically designed to make the perfect paella. They're wide, round and shallow to allow the rice to cook in a flat layer so it absorbs more liquid and retains more flavor. Some of these pans are made with cast-iron, but we prefer the lighter carbon steel pans because they heat up and cool down more quickly.
- **Pressure cooker:** A pressure cooker has a sealing lid that traps all the liquid and steam inside the pot. The pot pressurizes as soon as the contents come to a boil, cooking the food inside up to 70% faster. Stovetop pressure cookers used to be kind of dangerous, but today's models are better rated for safety.
- **Roasting pan:** A roasting pan is great for making a Thanksgiving turkey, but it has a variety of other uses, like cooking large quantities of vegetables or lasagna for a crowd. These large, rectangular dishes are oven-safe and look like an oversized casserole dish or a tall-sided baking sheet. You can use them with the rack for cooking large cuts of meat, or ditch the rack for cooking foods like roasted vegetables.
- **Saucepan:** Unlike braziers, Dutch ovens or stockpots, saucepans contain long handles and are sized significantly smaller. They come in a variety of materials, but we like try-ply or multi-clad stainless steel best for even heating and heat retention. It's good to have a 1-quart saucepan for heating small amounts of sauce (like barbecue sauce for a cookout), and a 2-quart saucepan is perfect for making rice or quinoa. A 3-quart saucepan isn't necessary if you have a Dutch oven, but it can be helpful for making small amounts of soup.
- **Saute pan:** A saute pan is similar to a frying pan, but it has tall, straight sides, and it always comes with a lid. The tall sides keep liquid from evaporating as quickly and also helps to prevent splatter. They're great for poaching eggs, pan frying or cooking large quantities of greens.

- **Omelet pans:** There open-topped and lid style omelet pans.
- **Skillet:** A skillet is essentially the same thing as a frying pan, but it has slightly taller edges. It also generally refers to a pan made with cast iron. These heavy-duty pans can be used on the stovetop or the oven. If you head out camping, you can even put it directly on the fire!
- **Grill pans:** A grill pan is designed to replicate grill cooking on the stove.
- **Stockpot:** These large pots are perfect for boiling water, making broth or stock, steaming vegetables or tamales (with a steamer insert) or cooking a large batch of soup. Unlike Dutch ovens, most stock pots are not intended for use in the oven, so we don't recommend them for braised dishes.
- **Tagine:** A tagine is an earthenware or clay cooking pot used for North African and Moroccan cuisine. It has a high lid that controls the way the steam escapes, forcing the condensation to drip down on the food as it cooks. You can make Moroccan chicken in a slow cooker, but it's more fun when you make it the authentic way!
- **Wok:** These small-bottomed, fluted-edge pans are typically used for Asian cuisine. They heat up quickly for stir-fried dishes, but you can also use them for steaming dumplings or deep-frying food. Carbon steel woks are the most traditional, but there are several options for stainless steel or nonstick woks.
- **Steamers:** A steamer pot is one with a metal screen that sits inside above the bottom. You place water on the bottom and the steam cooks whatever you put in the steam insert. This is great for steaming vegetables. Steamed broccoli this way is one of my favorites.
- **Pasta pots:** A pasta pot is designed to cook pasta. There are two main types of pasta pots. The first is one with a perforated locking lid so that you can pour the water out and be left with cooked pasta.
- **Double boilers:** A double boiler is like a steamer, except the insert is a pot with no holes. These are good for for food that cooks best with steam heat such as melting chocolate.
- **Milk warmer:** Most kitchens won't get a milk warmer, but if you like warm milk and prefer not doing it in a small sauce pot, you can actually buy a pot designed for warming milk on the stove.

Exercise-1

Draw different types of pots and pans:

Conclusion: ____

____

____

____

____

____

Remarks: ____

____

____

____

____

____

Signature

14

Frying Pans

Objective: Performance testing of three different types of frying pans (iron pan, stainless steel and nonstick pan).

Method: To check the performance of these pans, prepare 3 besan chillas on each type of pan by using equal quantity of oil and batter. Then compare and evaluate prepared dishes on 5 parameters i.e. Appearance, Texture, Aroma, Color and Taste by using 9-point hedonic scale.

Materials Required: 3 types of pans, materials for preparing besan chilla.

Frying Pan

A frying pan, frypan, or skillet is a flat-bottomed pan used for frying, searing, and browning foods. Frying pans come in different shapes and sizes. It is typically 20 to 30 cm (8 to 12 in) in diameter with relatively low sides that flare outwards, a long handle, and no lid. Traditionally, frying pans were made of cast iron, carbon steel or copper lined with tin, for their different qualities and properties.

Cast Iron

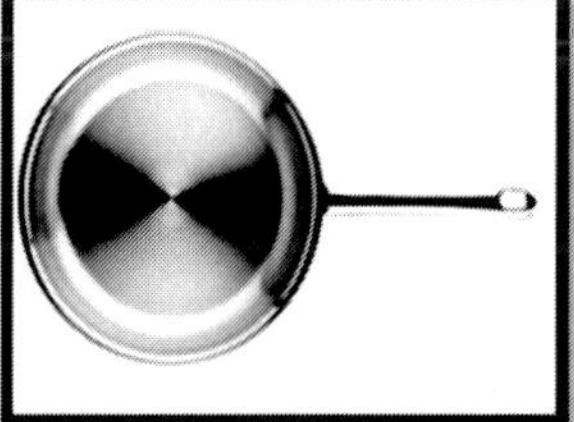

Stainless Steel

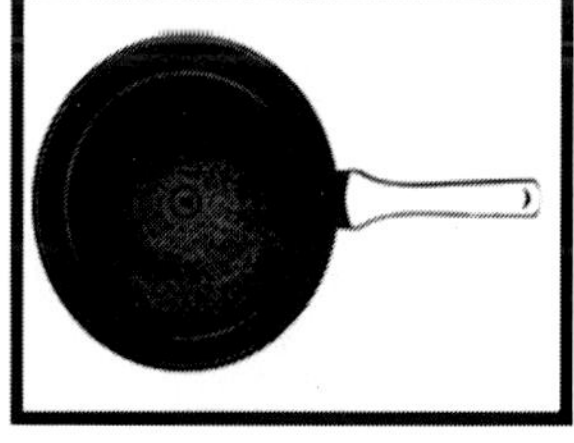

Non-Stick

Nowadays, most frying pans are now made from metals such as cast iron or stainless steel. The materials and construction method used in modern frying pans vary greatly. A coating is sometimes applied to the surface of the pan to make it non-stick. Every type of cooking pan material has its pros and cons and it's essential to consider several factors to determine which type will meet all of your at-home cooking needs.

The 9-point Hedonic Scale

The most widely used scale for measuring food acceptability is the 9-point hedonic scale. David Peryam and colleagues developed the scale at the Quartermaster Food and Container Institute of the U.S. Armed Forces, for the purpose of measuring the food preferences of soldiers1. The scale was quickly adopted by the food industry, and now is used not just for measuring the acceptability of foods and beverages, but also of personal care products, household products, and cosmetics. In the hedonic scale method the stimuli (actual samples or food names) are presented singly and are rated on a 9-point scale where the 9 categories range from 'like extremely' to 'dislike extremely'.

- 1- Like Extremely
- 2- Like Very Much
- 3- Like Moderately
- 4- Like Slightly
- 5- Neither Like nor Dislike
- 6- Dislike Slightly
- 7- Dislike Moderately
- 8- Dislike Very Much
- 9- Dislike Extremely

Exercise-1

Prepare 3 Besan chillas on iron pan, stainless steel and nonstick pan for testing. Explain the process of preparing the Besan chilla. Taste the preparedd chillas and evaluate each pan chilla on the 9-point Hedonic Scale.

Recipe of Besan Chilla

____

____

____

____

____

____

____

Testing of Prepared Dishes on Hedonic Scale

SN.	Name	Appearance	Texture	Aroma	Color	Taste
Cast iron						
Stainless steel						
Non-stick						

Results

- Best Appearance:
- Best Texture:
- Best Aroma:
- Best Color:
- Best Taste:

Remarks:

Signature

15

Toaster and Sandwich Maker

Objective: Operation and functioning of toaster and sandwich maker.

Method: To check the operation and functioning of toaster and sandwich maker, prepare sandwich on each brand of toaster and sandwich maker. Then compare and evaluate prepared sandwich on 5 parameters i.e. Appearance, Texture, Aroma, Color and Taste by using 9-point hedonic scale.

Materials Required: Two different brands of toaster and sandwich maker, materials for preparing sandwich.

Toaster

The toaster is a kind of equipment used to toast bread in order to make it crisp. It has two slots which can hold two or four slices of bread. The electrical element produces the heat needed to make the bread hot and crisp. The toaster is a little apparatus intended to toast numerous sorts of bread items. Created in Scotland in 1893 and persistently created from that point forward, the most well-known family unit toasting machines are presently the pop-up toaster and the toaster broiler. Pop-up toasters make toast from bread in 1 to 3 minutes by utilizing warmth components. Since the 2000s, pop-up toasters with more extensive openings have been produced, empowering these toasters to toast bagels and English biscuits cut down the middle. Toaster stoves have heat components above and underneath a barbecuing territory and capacity the same as a little scale customary stove. They are sufficiently extensive to warmth up such things as a cut of pizza or a burrito.

Construction

Like irons, toasters too may be automatic or non-automatic. The automatic models have a heat regulator which can be set at the desired temperature. At that temperature, the bread slices are released automatically and pop out of the toaster. In the non-automatic types you have to release the slices by pressing a knob. The main advantage of the automatic model is that you are free to do something else after putting in the bread slices as you don't have to watch over them constantly. There is no fear of their burning, either.You can also toast the bread while the toaster is kept on the gas. But in this type of toaster, you have to change the side of the toaster when one side is done.

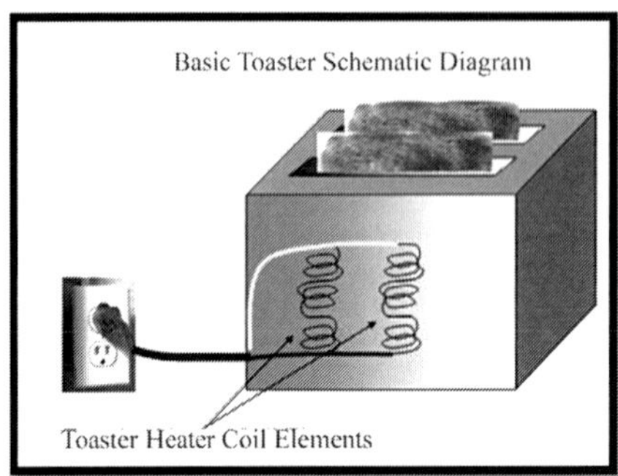

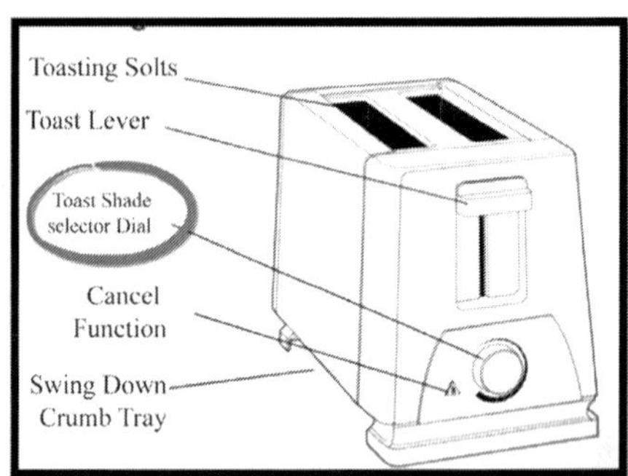

Toaster Operation

1. Operate the toaster twice on a light setting with no bread to disperse the initial element burning smell. Make sure the room is well ventilated. Once this step has been completed, do not use the toaster without bread slices occupying toaster slots as overheating may occur.

2. Place two slices of bread in the bread slots and press the handle down until it latches. The bread will be lowered into the toaster and the toasting cycle will begin.

   Note: The handle will not latch if the unit is not connected to the mains, and may not latch if the browning control is set to its minimum position

3. If the toast starts to burn or smoke before it has popped up, toasting can be stopped by pressing the 'Cancel' button.

4. If you should find that the toast is not dark enough you may wish to toast the bread again. Turn the browning control down and keep a careful eye on the toaster to ensure that the toast does not burn. Remember to increase the browning control setting for subsequent operations. If the toast is too dark, reduce the browning control before toasting the next slice.

   Note: Remember that dry or thin slices will toast quicker than moist or thick slices. Using a lower browning control setting for thinner slices is recommended. When toasting two slices at the same time, try to use the same size and consistency of bread to achieve equal browning.

5. When toasting has finished the handle will rapidly rise to the fully raised position. The toasted bread can now be removed and the toaster is ready to recommence toasting. Should the toast not be accessible from the standard pop up position, the extra lift function may be used. Simply push the lever upwards to activate the high lift function and remove toasted items.
6. Avoid using the toaster on the maximum browning control setting. Most breads toast well below the 4 setting
7. When toasting smaller slices of bread, use of the high lift lever function may be required. Once the toasting cycle has finished and the lever has popped up to its standard preset position, simply push the lever further upwards for additional lift and extract your toasted items.

Precautions While Using

- Do not use a fork to remove the toast. Because you may get an electric shock since the coils are not insulated.
- Do not place the bread with filling in the toaster. Because the filling will melt and catch fire.
- Do not drop or shake for this may damage the heating element.
- Before cleaning, disconnect the toaster and food processor from the electric supply.
- Never cover the toaster with anything. This will cause the toaster to overheat, causing irreparable damage and may interfere with the automatic on/off functions. If covered, the toaster and/or items covering the toaster, may ignite.
- Bread can burn. Do not use the toaster near flammable materials. Do not place the toaster close to curtains or other materials that could easily catch fire. The toaster should not be left unattended when in operation.
- Never force bread into the toasting slots. Never toast slices of bread that have been buttered. Take extra care when toasting items that contain sugar, jam, preserves, raisins or other bits of fruit, as there is a risk of overheating. Don't use bent, damaged or broken bread slices in your toaster as it may get jammed in the slot.

Cleaning and Care

Ensure the power is set to OFF and the unit has been allowed to cool. Be sure to remove the mains plug from the power outlet before proceeding to clean your toaster

- The exterior of your toaster may be cleaned with a soft damp cloth and polished with a dry soft cloth if necessary. Do not use abrasive cloths or cleaning agents.
- Dispose of crumbs by sliding out the removable crumb tray and emptying. Regular cleaning of the crumb tray is required as accumulation of crumbs may pose a fire hazard
- Additional cleaning and/or disposal of crumbs may be achieved by turning the appliance upside down over the sink whilst gently shaking. Any additional debris or crumbs should fall out and can be washed down the sink.

Sandwich Maker

A sandwich maker is an electronic device used to prepare closed sandwiches. In addition to sandwiches, the compact appliance can cook waffles, puffs, and pancakes. A sandwich maker is primarily characterized by its two heating plates that perform the role of clamping the sandwich once inserted. Conventional sandwich makers usually prepare two to four sandwiches. But due to its versatility, it can also perform the role of a toaster and brown slices of bread.

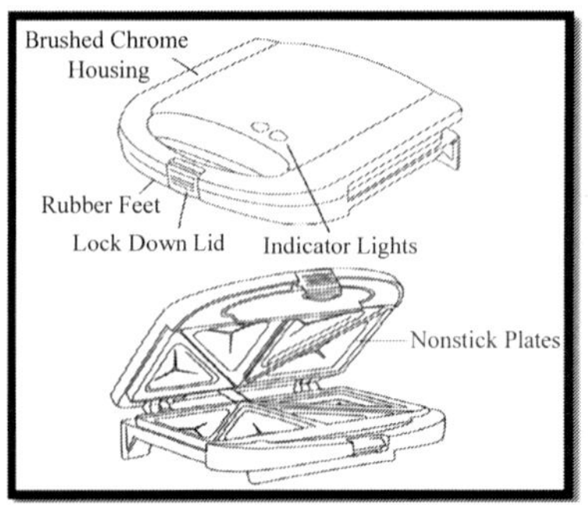

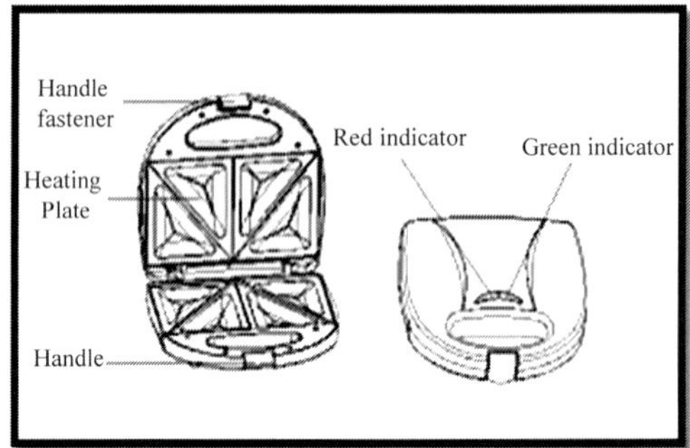

Maintenance and Cleaning

- Do not disassemble your sandwich maker for cleaning.
- Never immerse your sandwich maker in water or place it in a dishwasher.
- Do not use anything abrasive (such as metal scouring pads or steel wool) that can scratch or damage the non-stick coating or exterior finish.
- Do not use metal utensils to remove food residue because they can damage the non-stick surface.

Exercise-1

Prepare sandwiches on toasters of three brads for testing. Explain the process of preparing the sandwich. Taste the prepared sandwich and evaluate each sandwich on the 9-point Hedonic Scale.

Recipe of Sandwich

____

____

____

____

Testing of Prepared Dishes on Hedonic Scale

SN.	Name	Appearance	Texture	Aroma	Color	Taste
Philips						
Borosil						
Bajaj						

Results

- **Best Appearance:**
- **Best Texture:**
- **Best Aroma:**
- **Best Color:**
- **Best Taste:**

Conclusion: ____

____

____

____

____

____

Remarks: ____

____

____

____

____

____

Signature

16

Vacuum Cleaner

Objective: To study parts, operation and functioning of vacuum cleaner.

Vacuum Cleaner

A vacuum cleaner, also known as a sweeper or hoover, is a device that uses an air pump (a centrifugal fan in all but some of the very oldest models), to create a partial vacuum to suck up dust and dirt from floors and from other surfaces such as upholstery and draperies.The dirt is collected by either a dustbag or a cyclone for later disposal. Vacuum cleaners, which are used in homes as well as in industry, exist in a variety of sizes and models—small battery-powered hand-held devices, wheeled canister models for home use, domestic central vacuum cleaners, huge stationary industrial appliances that can handle several hundred litres of dust before being emptied, and self-propelled vacuum trucks for recovery of large spills or removal of contaminated soil. Specialized shop vacuums can be used to suck up both dust and liquids.

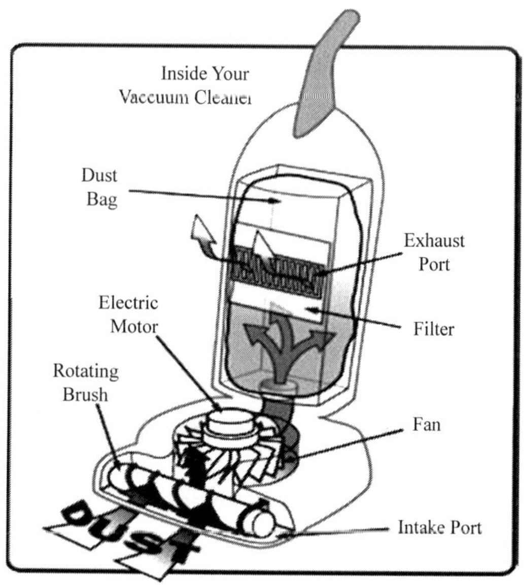

Construction and Working Principle

A vacuum cleaner is one of the most essential home appliances one can have for cleaning/sanitizing one's home. It works on the principle of flow of air from area of high pressure to area of low pressure. An electric motor is attached to a fan that spins it at high velocities. The fast spinning fan creates a region of low pressure inside the suction hose of the vacuum cleaner. Air, along with dust and debris is sucked into the suction hose because of the pressure difference between outside and inside of the suction hose.Once inside the suction hose, air and debris travels towards a dust collecting bag which is made of porous woven cloth or towards a canister, depending upon the type of vacuum cleaner. The dust and debris gets accumulated in the collecting bag and air, relatively cleaner flows out through the exhaust port after passing through a HEPA filter which further sanitizes it and removes the allergens from it.

Types of Vacuum Cleaners Based on Their Use

- **Canister Vacuums:** Canister Vacuums are normally equipped with a suction hose and retractable cord to make maneuvering easy. They are normally used on hard floors, or it's used as a vacuum cleaner for sofa and cars. They are highly known for their flexibility, motor control, and versatility. They are the best, when cleaning of furniture and multi surfaces are approached. They are one of the most traditional and commonly bought vacuum cleaners, since they can be used in many ways and are highly reliable. They are the best vacuum for hardwood floors.
- **Deep Cleaner or Carpet Cleaner:** Otherwise known as carpet extractors or shampoo shampooers, they are highly used as vacuum cleaners for carpets. They first spray the carpets with foamy, soap water and then use vacuum technology to suck up water droplets and soapy discards, and it also sucks up fibers and dust particles along with them. They are not designed for hardwood floors and they are not commonly used in daily lives, as they are used on rare cleaning occasions. This is one of the unique ones, since they sprinkle and clean. They are the best vacuum cleaners for carpets
- **Upright Vacuum:** Upright Vacuum Cleaners are preferred due to their versatile usage and its adjustable features and an innovative motor. They can be used as vacuum cleaners for hardwood floors and also carpets. They have a beater brush that is powered by an electrical motor and loosens up the dirt and sucks it up in its suction pump. Some Uprights might contain a bag to collect dust, while some might not. They are of

adjustable height and adjustable features that help us to change the mode from hardwood floors to carpets.

- **Handheld Vacuum:** (Vacuum Cleaner for Cars): Portable, Cordless, and Mediated, Handheld Vacuum are highly used to clean after pets. It is one of the vacuums used to clean dog hair. They clean the sofa, cars and anything of furniture. They are comfortable, chargeable and are used to clean small materials. They are the best vacuums to clean sofas and other furniture too.
- **Pet Vacuum:** As its name denotes, Pet Vacuum is used to clean and absorb pet hairs and clean the place, making it allergy free. The Beater Brush loosens the hairs and makes it easy for the vacuum to absorb through the carpets and floors. They are the best vacuum cleaner for dog hairs. They can also be used as upholstery cleaners to clean the little furniture, where pets tend to shed hair.
- **Robotic Vacuum:** Robotic Vacuum, which was created based on the latest vacuum technology, is one of the most convenient appliances created in recent times. Though being not as traditional as its forefathers, Robotic Vacuums have inbuilt sensors that can be controlled using controls, and do not require manual support. They are the best vacuums for laminate floors. You can practically operate the vacuum from exactly where you are, using the controls provided. They are not as versatile as others, yet, these vacuums can be used on hardwood floors and carpets.
- **Sweeper or Stick Vacuum:** Being a fusion between upright vacuum and handheld vacuum, Sweeper or Stick Vacuums are normally used to clean kitchen and bathroom floors. They are manually operated, and instead of bags to collect dirt, they possess cups to collect dust sucked from the floor. They are easily handled and are battery charged. They are the best vacuum for stairs and floors that need to be cleaned in angles. They can be used like upright vacuum, yet being handled like a handheld one. They are lightweight and used for cleaning quick mishandles and dust issues. While Handheld Vacuums are used to clean sofas and cars, other old fashioned vacuums can be used to clean floors and carpets. It is necessary that the vacuum has sufficient wattage. Normally, a corded vacuum requires 1000 to 2000 watts, while cordless requires 20 to 200 watts.
- **Bagged and Bagless Vacuums:** Bagless Vacuums are mostly preferred than bagged ones since they save a lot of money without the replaceable bags. Yet they also require periodic cleaning for the HEPA filters like the bagged ones. The Filters are more often replaced in the bagless vacuums than in that of bagged ones.

Safety Checklist When Using A Vacuum Cleaner

- Unplug the vacuum cleaner whenever the appliance is not in use. This habit prevents accidents and also protects the cord.
- Use only a wet and dry vacuum cleaner to clean wet surfaces. Using any other appliance will damage the motor and increase the risk of electric shocks.
- Don't drop the vacuum cleaner. In case you drop the appliance, call a technician to check the appliance before using
- Carefully store the cord of the vacuum cleaner in a safe place to avoid damage.
- Avoid cleaning toxic or flammable materials with a vacuum cleaner.
- Avoid picking up sharp objects such as nails, screws, and coins with a vacuum cleaner
- Store the vacuum cleaner in a cool and dry place.

Testing of Performance of Vacuum Cleaner

S.No	Name	Regulate movement	Cord length	Easy to clean	Suction power	For deep cleaning	Noise level
Agaro							
1							
2							
3							
Eureka Forbes							
4							
5							
6							
Philips							
7							
8							
9							

Results:

Conclusion:

Remarks:

Signature

17

Microwave Oven

Objective: To study the parts, functioning and performance of microwave oven.

Introduction

A microwave oven, commonly referred to as a microwave, is a kitchen appliance that heats and cooks food by exposing it to electromagnetic radiation in the microwave spectrum.

Main Components of Microwave

A microwave oven is used to cook (or heat) food with the help of microwaves produced by magnetron – the device producing microwaves in the oven. Microwave ovens are so quick and efficient because they channel heat energy directly to the molecules (tiny particles) inside food. Here are several parts of a microwave, and they each have a specific role in heating the food. They consist of internal components (Magnetrons, high-voltage transformer, waveguide, cooling fan, turntable, and cooking cavity) that help cook the food and external pieces (Case, power cord, door, and control panel) that prevent radiation from escaping. In a nutshell, these include:

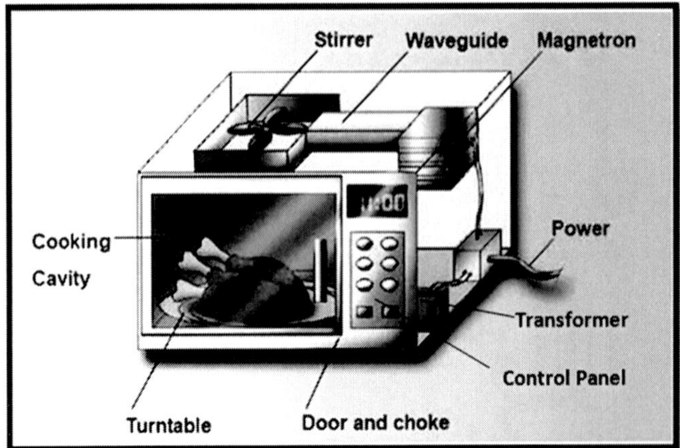

- **Transformer:** Microwaves require a high voltage of power to modify the power level and give a sufficient voltage to the machine. This is the place where the high voltage transformer comes in, as it can change the voltage over to a lot more significant level. Most microwaves require at least 3000-4000 volts to heat food. Standard outlets use about 115 volts of power. The high-voltage transformer converts that lower voltage into higher energy to cook food. As with any part, the size and power level will vary depending on the type of microwave you have.

Transformer

Magnetron

The Magnetron: The microwave's magnetron consists of a vacuum-like tube and a strong magnet and sits on the same side as the control panel. This piece of a microwave gets the high voltage of power provided by the transformer and converts it into microwave energy to warm the food. Like a human brain sends messages to the body, magnetrons produce microwaves that travel through the waveguide and bounce from the oven's metal interior to the water molecules in the food. The tiny waves cause the water molecules to vibrate. The friction of the vibrating molecules produces heat, thereby cooking the food.

- **A waveguide:** The waveguide is a hollow metal tube that extends from the magnetron to the cooling fan. It directs the flow of microwaves from the magnetron to the cooking chamber. While these waves transfer, they reflect off the inner walls and into the food, cooking it. A waveguide includes a thin plastic or cardboard-like cover to protect the internal components from food and splatters during cooking.

Transformer

Magnetron

Cooling fan or stirrer: The cooling fan in older versions is a part of the microwave ceiling, with a covering to protect it against splatters. The fan prevents the unit from overheating. If too many heat waves build up, not only will the food possibly burn but the appliance may become too hot (more than 150 degrees) and break down. When the microwave has been in use for a long time, it gets hot inside the chamber. The cooling fan knows when to turn on and off. It automatically kicks on and begins to cool your appliance. However, more seasoned microwaves didn't have cooling fans yet rather highlighted a stirrer fan which was intended to do exactly the same thing. Although, the stirrer would be fitted at the highest point of the microwave inside the cooking pit and attempted to flow the microwaves to help guarantee in any event, warming of the food and forestall overheating.

Cooling Fan

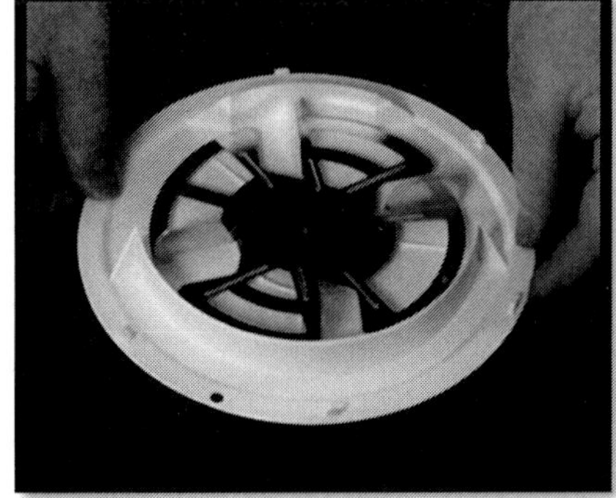

Stirrer

The rotating disc/ Turntable: The turntable is a glass plate inside the microwave hole, which is normally removable to make for simple cleaning. The motivation behind the turntable is to pivot the food around so it cooks all the more equally, staying away from hot or cold spots. The turntable openings onto the lower part of the microwave pit utilizing edges, permitting it to effortlessly secure set up. You place your food on a round glass or plastic plate so that the food will rotate during cooking. The purpose of rotating food is that heat will evenly distribute through and cook it thoroughly.

Cooling Cavity

Turntable

- **Cooking Case or Cavity:** The cooking case or cavity is indeed one of the most important microwave components in order to achieve an efficient heating. It is the deep inner compartment that holds the food while it cooks. The inner walls consist of thick, stainless steel to absorb the waves when they leave the food.
- **Exterior Case:** The outer case is the outer metal shell of the microwave that houses the inner components and cooking chamber. Waves stay inside the cavity when the sealed door creates a seal to protect you from radiation during the cooking cycle.
- **Door:** The door consists of a window to let you view your food. Some microwaves have a light that comes on during heating so that you can see clearer. A metal mesh lining between the glass layers prevents minute waves from leaking out. This allows you to see in but keeps radiation away. The rubber seal around the door blocks radiation from leaking out. To ensure safety, the seal must be in good working condition. The presence of cracks could indicate wear and tear.
- **Control Panel or Inspection Window:** This is the prime console from where the user gives the instructions to cook or warm the food. The

design of the microwave control panel varies from model to model. In some variants, the interface is touch screen, while the rest have buttons to control everything. It allows you to choose your settings. Use a knob or buttons to determine the cooking time, cooking power, or mode, depending on what model you have. Most appliances include a defrost setting, a timer, and a clock. Some units may include an option to pre-set cooking times. All microwave control panels have a door release button. You may have extra selections for potato, popcorn, or other types of food.

- **Power Cord:** This component speaks for itself. It connects the microwave to electricity. You can detach the power cord from the back of the unit if you need to replace it.

Functioning of Microwave

Microwave ovens work on the principle of conversion of electromagnetic energy into thermal energy. Electromagnetic (EM) energy refers to the radiation (waves) comprising an electrical field and magnetic field oscillating perpendicular to each other. The process of heating food in the microwave oven is fairly simple; however, the mechanism involved in that process is somewhat atypical. After the generation of microwaves at the magnetron, they are guided by the waveguide towards the food inside the cavity. The microwaves penetrate through the surface of the food and reach the water molecules present inside it. As the orientation of the electric field changes over time, the polar molecules of water attempt to follow the field by changing their orientation inside the material to line up along the field lines in an energetically favourable configuration (namely, with the positive side pointing in the same direction as the field lines). As these molecules change direction rapidly (millions of times per second at least), they gain energy, which increases the temperature of the material. This process is called dielectric heating. The microwave energy diminishes according to the inverse square law, and therefore, the cavity chamber, where we place food, is designed in such a way that it carries out the maximum efficiency of the heating effect of microwaves. Furthermore, most of the microwave ovens come with a door switch that does not allow the process to initiate until the door is completely sealed.

Special Features of Microwave Oven

- **Automatic programming:** This is an automatic sequence of different power levels and cooking times. More than one setting can be programmed at once, so that cooking sequences can be carried out one after the other. It can be programmed to defrost food and then automatically switch to a cooking setting. It can be made to start cooking at a higher setting, then to complete cooking at a lower one, or vice versa.
- **Browning Element:** This is electrical element mounted at the top of the oven cavity for the purpose of top browning or crisping food. It works just like a conventional grill. During this operation, the fan in the oven operates but no microwave energy enters the oven. Dry radiant heat from the Browning Element adds colour and crispness of the food after it has been cooked.
- **Defrost Control:** Defrosting in the microwave is a speedy, hygienic and healthy process. When the defrost control is used along with the timer, the microwave energy in the oven is automatically cycled on and off to allow rest periods in between short bursts of microwave energy. These short gentle bursts ensure that thawing is carried out at a uniform rate so that food thaws without cooking. In a microwave oven, you can simply defrost, and then cook in one simple operation for perfect results, at a moment's notice. This feature automatically selects 30 percent power level, which is the ideal power level for defrosting. However, it is also possible to defrost by selecting a lower power level.
- **Tips to speed defrosting**
  - Flex pouches and packages which cannot be broken up or stirred. This distributes heat more evenly.

    - Cover meat with wax paper. It retains warmth around the food as it begins to defrost.
    - Shield thin edges of fish and poultry to prevent them from cooking.
    - Pour off liquid from poultry frozen in a plastic bag. It absorbs heat and slows the process.
    - Observe standing time to completely thaw food especially the thicker, innermost parts.

- **Reheating:** Unlike as in conventional cooking, reheating in the microwave is quick and efficient. Reheat food without any loss of quality or texture, tasting freshly cooked, not just warmed over.
- **Some Reheating Tips**
    - Main dishes taken out of the refrigerator are best reheated at 50 per cent should be reheated at a lower power setting.
    - Test by feeling bottom of the plate being used. If the food is hot the plate will be just warm. When reheating in a plate place thick or dense food towards the outside and delicate foods in the centre.
    - Spread out food when possible for quick, uniform heating.
    - Cover main dishes for effective reheating. Plates can be covered with wax paper or a plastic wrap. Stir main dishes, casseroles and gravy dishes to distribute heat.
    - Add sauce or gravy to dry meats to provide moisture. If sauce is not suitable or top your taste, wrap in wax paper.
    - Reheat desserts with filling, carefully. The pastry may be comparatively cool while sugary fillings will be very hot.
    - Fibrous vegetables such as asparagus or broccoli lose texture when reheated. Bury these in sauces or gravies. Moist or starchy vegetables like potato taste fresh when reheated well wrapped.

Factors Affecting Microwave Cooking

These food characteristics after all form of cooking, but microwave speed makes the difference more pronounced.

- **Starting Temperature:** The colder the food the longer it will take to heat up or cook and vice versa. Remember too, that room temperature may differ appreciable with season and that cooking times may be longer on a cold winter day.

- **Density:** Dense, heavy food like meat or vegetables may longer to cook than light, porous foods such as bread, cakes and puddings, etc.
- **Quantity:** Small amount of foods take less time to cook than larger quantity. If the amount of food being cooked is double is double, the cooking time will increase by about half as much time.
- **Size:** Small pieces of food cook faster than large ones. Since microwaves penetrate food ¾ to 1½ inches, uniform pieces under 2 inches in diameter cook faster and more evenly.
- **Shape:** Thin parts of uneven foods cook faster than thick parts, while uniformly thick foods cook evenly. Thin parts of food should be placed towards the centre of the dish where they receive less microwave energy. Shield thin parts with small pieces of aluminum foil to slow down cooking.
- **Amount of Fat and Bone:** Bone conducts heat so the side of meat with bone cooks first. Boneless cuts cook less rapidly but more evenly.
- **Composition of Food:** Fats and sugars absorb microwave energy faster than other liquids and so they will cook faster. Evenly distributed fat tenderizes and helps cook food evenly.
- **Sensitive Ingredients:** Ingredients like cheese, eggs, cream, condensed milk etc. attract microwave energy. These overcook very quickly and can toughen or curdle. Use lower power setting or bury them in less vulnerable foods during cooking.

Precautions While Using Microwave Oven

- Like many other electrical appliances, it is important to follow the manufacturer's instruction manual for recommended operating procedures and safety precautions for your oven model.
- Use microwave-safe cookware specially manufactured for use in the microwave oven. The range of dishes that can be used for microwaving is much wider than that used for conventional cooking. Remember that dishes should not act as a barrier to microwave energy. Metal dishes are unsuitable. Some microwave manufactures allow the use of aluminum cookware. Before using a utensil for the first time, carry out this test; place the utensil in the oven cavity. Place a glass tumbler with water inside. Microwave on HIGH for two minutes. If the utensil becomes warmer than the glass of water, it should not be used.

- **Paper:** Paper plates, absorbent kitchen paper napkins, paper cups and cartons all have their uses. Paper is suitable only for reheating, not for cooking. (Do not use recycled paper such as newspaper).
- **China, Glass and Pottery:** Oven proof china glass ('Borosilware') and pottery dishes are perfectly suitable. Check that there are no metal rims or glued-on handles. Ceramic dishes should be non-porous, heat it in the microwave for 15-20 seconds. The dish should not feel warm.
- **Plastic:** Plastics like Styrofoam are safe heating but melt or distort at cooking temperatures. Melamine, however, is not recommended as it may charr.
- **Cotton and Linen:** 100% cotton or linen napkin can be used for short reheating purpose, like warming bread rolls or croissants.
- **Straw and Wood:** Use these only short reheating times. Wooden spoons may be left in the microwave for short cooking periods.

- A microwave oven should not be operated with an open, bent, or broken door.
- The liquids should not be heated longer than the recommended temperature as it can cause the water vapors to reach the electric components and interfere with their working.
- To avoid any mishappening, it is recommended not to stand directly in the front of a microwave oven while it is operating.
- It is essential to periodically clean the cavity with water and mild detergent. It is recommended to not use scouring pads, steel wool, or other abrasives for the cleaning process.

Advantages of Microwave Oven

- The volumetric heating process of microwaves is their most prominent characteristic. In the conventional cooking method, the heat must spread inwards from the surface of the food item, whereas the spread of heat in the case of microwave oven is done in a controlled manner with the help of the microwaves.
- It's a quick and convenient method of heating food and leftovers.
- Since microwaves can only interact with polar substances like water, they cannot affect the nutritional value of those ingredients that are non-polar. Other conventional cooking methods, however, may destroy some polar as well as non-polar ingredients during the process.

- The user interface and micro-controller facilitate precise control over the cooking temperature.
- The ease of the cooking process in a microwave oven also results in easier cleaning of the equipment after use.

Disadvantages of Microwave Oven

- It is important to take care of what kinds of utensils are being used in a microwave. A dish that is not microwave-safe will set off a chemical reaction between the food and the container.
- The cost of equipment is high in comparison to other conventional cooking methods.
- Microwave leakage may lead to electromagnetic interference with other electrical equipment present in the surrounding vicinity. The pacemakers installed in some patients are particularly vulnerable to such radiation leakage.
- Microwave radiation can heat body tissue the same way it heats food. Exposure to high levels of microwaves can cause a painful burn. In particular, the eyes and the testes are vulnerable to microwave heating because there is relatively little blood flow in them to carry away excess heat.

Conclusion

Microwave ovens are common in households and are established there as devices of everyday use. In today's modern world, every family needs an easy way of working and time-saving method for any task so that they can divert to the other activities once they are finished with it. Women in general who are bored with their kitchen work and daily routine can enjoy making food in an easy way with more taste within the fraction of a second using a microwave oven. However, we should not forget to follow every safety measures provided in the guideline.

Exercise-1

Answer the following questions:

1. How do microwaves cook the food?

2. Are microwave ovens safe?

Exercise-2

To understand the use of different parts and functioning and to check the performance of a microwave oven, prepare a dish in it. Taste the prepared dish; evaluate the dish on the following parameters and give rating to each parameter on 9-point hedonic scale.

Rating	Parameters for evaluating the prepared dish				
	Appearance	Texture	Aroma	Color	Taste
1-Like Extremely					
2-Like Very Much					
3-Like Moderately					
4-Like Slightly					
5-Neither Like nor Dislike					
6-Dislike Slightly					
7-Dislike Moderately					
8-Dislike Very Much					
9-Dislike Extremely					

Results: ____

____

____

____

____

____

Conclusion: ____

____

____

____

____

____

Remarks: ____

____

____

____

____

____

Signature

18

Mixer/Juicer

Objective: To study the construction, functioning and working principle of mixer/juicer in preparation of juice.

Material Required: Mixie/juicer, fruits (according to season), knife, spoons

Mixer or Mixie

The domestic electric food juicer/mixer/grinder has become an integral fixture in the typical modern kitchen. Generally functional as liquidizers, grinders and juicers, this paraphernalia comes in handy for grinding/pulverizing/powdering dry foodstuffs either raw or roasted (cereals, grains, masalas, coffee seeds, etc.), for mixing liquids and for converting food— with or without the presence of water (or vegetable oils)—into forms of slurry or pulps. It is also used for extracting juice from fruits or vegetables. Basically, whether the purpose is to whip up chunky chutney or blend a cool lassi. The mixer grinder is equipment which saves a lot of our time and labour. It can grind dry masala and make chutney or even lassi very quickly.

Mixer Grinders

Food Processor

The mixer has two basic functions:

a) **Dry grinding** – this is to do dry grinding of masalas, coffee seeds, cereals like dalia etc.

b) **Wet grinding** – this operation needs some amount of liquid like making chutneys, milk shakes, lassi, masala paste etc.

Construction

Mixer consists of a base which holds the motor and controls, a glass, plastic or steel food container that fits on the base and a cover for the container. Inside of the container, there are many sharp steel blades. Most models available today have many advanced functions of a food processor like kneading atta, slicing, shredding, mincing, chopping, etc. The cost of a mixie or food processor varies with the number of attachments and functions.

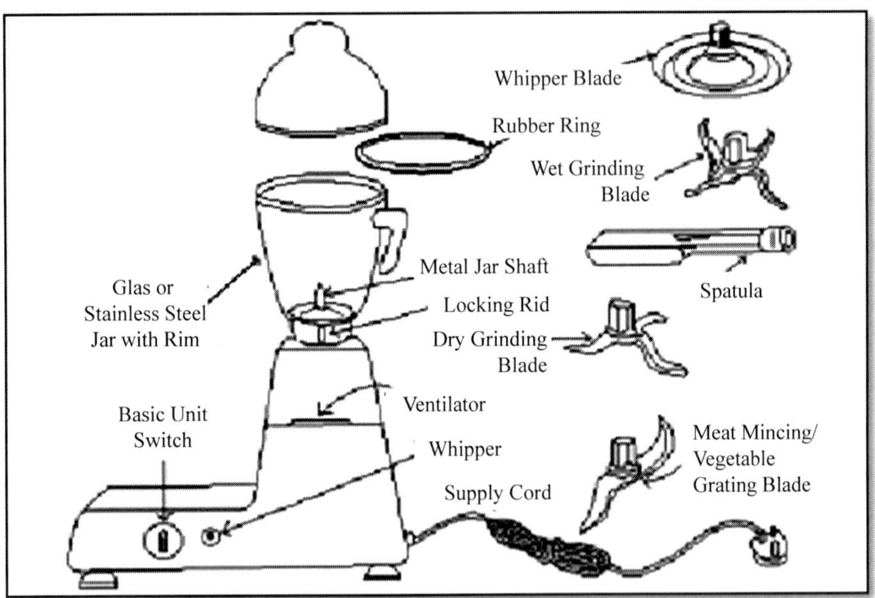

Different Parts of a Mixie

Comparative tests to be performed on various mixie/grinder/juicer are

a) **Grinding coffee** - As per IS, the recommended quantity of roasted coffee beans is taken for grinding and filtered in the three sieves of different microns. The material retained on each of the first two sieves shall not be more than 20 per cent of the weight obtained at the end of the test. The material passing through the third sieve shall not be less than 30 per cent of the same weight.

b) **Whisking egg white** - In this test, the material shall remain in the bowl when inverted for five seconds.

c) **Making idli batter**- This test indicates uniform performance in making a batter of rice and black gram. This requires grinding and mixing of the material in water to make a uniform, smooth and homogeneous

d) **Batter for further cooking** - The mixture shall be smooth and frothy, and no lumps shall be detected.

e) **Grinding rice**: The material retained on 1.4 mm sieve shall be 10 per cent maximum; on 1.0 mm sieve, 15 per cent maximum; and on 0.5 mm sieve, 70 per cent maximum.

f) **Juice extraction**: Carrots can be chosen and operated as per the guidelines of the manual. The observation was mainly for smooth and uniform/ homogeneous juice. Juice pouring out of the vent should have an even flow. The rate of juice extraction shall not be less than 500 gm/ minute and the amount of juice extracted shall not be less than 65 per cent. Finally, the juicer shall be dismantled, cleaned and inspected. There shall be no spillage/leakage.

Precautions While Using

- Use the mixer with three-pin plug and at the correct wattage and voltage as indicated. Why is this necessary? This is necessary to avoid overheating of the motor.
- Cover the blades with enough food in the bowl; otherwise, the grinding will not be adequate.
- While the food should be enough to cover the blades, it should not fill more than ¾ of the bowl/jar; otherwise, the motor will be strained and may burn out.
- If you are using ice, use crushed ice instead of whole pieces. As big pieces may break the blades.
- For longer life of the mixer, do not run the motor continuously for long periods. Always give break after some time and increase or decrease the speed gradually.
- For removing sticky food clean by adding warm water, some detergent and turning on the blender.

Care and Maintenance

Here are some maintenance tips you should know about how to use juicer mixer grinder:

1. **Clean the mixer grinder after every use:** As the mixer grinder is quite regularly used in the kitchen, make sure you clean it after every use. Clean all the jars with dishwashing liquid and wash them with water. If you made any item that has a strong odour, use lukewarm water to rinse

them. Wipe the jars with a microfiber cloth and put them upside down to let them thoroughly dry. To clean the body of the mixer grinder, use a toothbrush to remove the dirt.

2. **Always lock the jars before every use:** To prevent any damage to the machine, make sure that all the jars are properly fitted on the machine before switching it on. Not locking the jars properly can damage the blades and adversely affect the working of the mixer grinder.
3. **Gradually increase the speed:** Instead of starting the mixer grinder at full speed, go from low to medium to high. This can help you a lot in increasing the overall life and performance of the machine.
4. **Don't use hot ingredients:** Before grinding, make sure the ingredients are at the room temperature. Allow the ingredients to cool down at least for 7 to 10 minutes before pouring them into the jar. And, avoid filling the jar to the brim. Too much content or too hot content may get splashed all over your kitchen.
4. **Don't use unrecommended accessories:** Always use accessories recommended by the manufacturers. Spare parts and accessories from a local store or some other brand may not fit your machine properly, leading to damage. If there's any problem with your mixer grinder, it is best to get in touch with the service centre.
5. **Allocate a dedicated space to the mixer grinder**: Ensuring that the mixer grinder is stored in a proper place. Make sure there are no objects that could fall on the machine and cause damage. To avoid electric shock, don't place the machine where there is water around. Also, it would be best to keep it out of reach of children.

Testing Parameters

Different brands will be tested according to the following parameters

1. **Construction** can cover the following criteria as internal wiring, components, supply connections and external flexible cables and cords, screws and connections, creepage distance and clearance and resistance. Besides this all brands should clear the test screws and connections creepage ,distance and clearance , resistance to rusting controls (controlling switches shall be capable of breaking the stalled motor current at the maximum rated voltage six times without failure) endurance (the food mixer shall withstand electric strength test)to heat and fire , resistance to rusting etc.

2. **For electrical and mechanical safety there should be** protection against electric shock, electric insulation and leakage current at operating temperature moisture resistance , insulation resistance and electric strength, stability and mechanical hazards, mechanical strength, temperature withstand test for bowl ,strength of assembly.

3. **Mechanical strength** The appliance shall have adequate mechanical strength and be constructed to withstand such rough handling that may be expected in normal use. Accessible parts of solid insulation shall have sufficient strength.

4. **Insulation resistance and electric strength**: This covers two aspects:

   a) Leakage current and electric strength at operating temperature

   b) High voltage: Appliance shall withstand 1,250 volts for one minute. No breakdown shall occur during the test.

5. **Wattage:** One of the first things to check when it comes to a mixer grinder is how much wattage does it have. Higher wattage is useful for grinding tougher ingredients and at a faster speed. Anything from a 500 W to 750 W is good enough for an Indian kitchen.

6. **Speed:** The RPM of a mixer grinder is also a deciding factor, as it tells you how fast the blade can rotate per minute. For a domestic mixer grinder, anything from 18000 to 23000 RPM is good enough. A juicer mixer grinder will ideally require low to medium RPM for working efficiently. At a higher RPM, liquids or spices may not be blended or grinded effectively. Look for quality mixer grinders that have speed control knob/switch to change the settings as per requirement.

7. **Jars:** The higher the number of jars, the better. Look for a mixer grinder that has 2 - 3 jars with varying capacities to meet your requirement. You can use one for grinding spices, one for making chutneys and another for juicing. A durable body and a break-resistant lid is a must-have to withstand the daily wear and tear in the kitchen. Also, check for an ergonomically designed handle for a good grip.

8. **Blades:** Blades are the ones which do all the work, so it definitely makes sense to verify their quality. Make sure that the material is stainless steel and preferably of 304 grade. The grade of steel matters, as it tells you whether the steel is resistant to oxidation and corrosion. High-quality steel blades also retain their sharpness for a longer period of time.

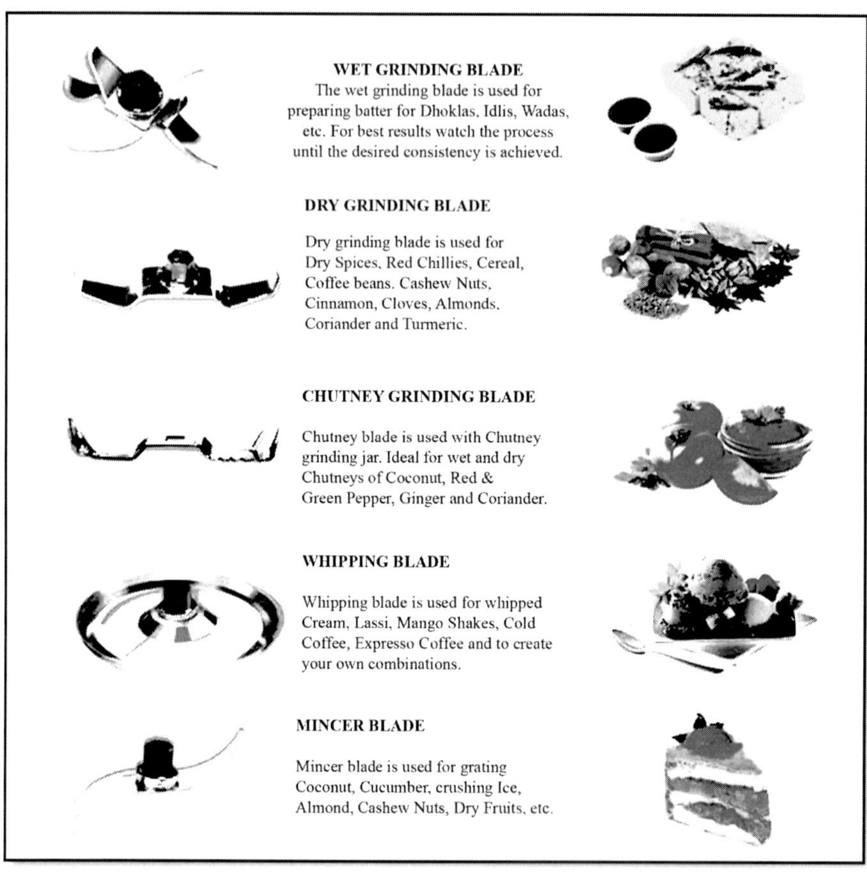

Different Types of Blades

9. **Warranty:** If the company offers you a good warranty on your product, then you can be rest assured that it is in fact a quality product. A warranty from 2 to 5 years is something that you should check for when selecting a model. Look for an additional warranty on the motor. This will ensure that your appliance is protected for a longer period of time.

10. **Price:** The price of your mixer grinder will depend upon the features it offers you as well as the overall quality. There are some inexpensive options in the market, but they may not have the best features.

Exercise-1

Select 10 brands of mixer /juicer and test for its on parameters construction, operation, performance, electrical and mechanical safety, and endurance, among others.

S.No.	Brand Name	Construction	Operation	Performance	Electrical and mechanical safety	Endurance

Exercise-2

Following are the six best brands of Juicer Mixer Grinder in India. List and rank accordingly:

6 Best Juice Mixer Grinders in India: Maharaja, Philips, Prestige, Bajaj, Havells, Inalsa, Morphy

Rank	Total Score out of 100 (Rounded off)	Brand	Model Rating (Watts)	Guarantee/ Warranty (Years)	MRP/ Retail Price (Rs)	Manufacturer/ Marketer

Results: ____________________

____

____

Conclusion: ____________________

____

____

Remarks: ____________________

____

____

Signature

19

Refrigerator

Objective: To study the components, functioning and performance of refrigerator.

Introduction

The refrigerator (called a 'fridge' by most people) is a box like equipment which uses electricity to keep the temperature inside the box lower than the atmosphere outside. Food kept at such a low temperature thus does not spoil for a long time. The refrigerator also serves the purpose of giving us cold water and ice in summer. By preserving food in the refrigerator, one can delay the growth of bacteria, keep food items fresh and thus save time and energy of the homemaker in meal preparation.

Components of A Refrierator

A refrigerator consists of several key components that play a crucial role in the cooling process:

- **Expansion valve:** Also referred to as the flow control device, an expansion device or capillary tube which controls the flow of the liquid refrigerant (also known as 'coolant') into the evaporator. It's actually a very small device that is sensitive to temperature changes of the refrigerant.

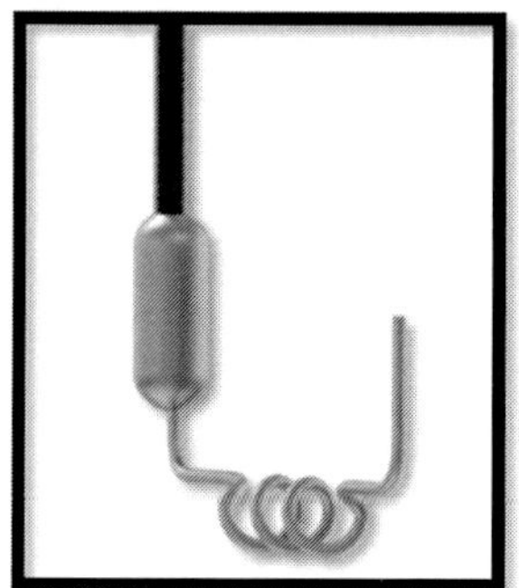

- **Compressor:** The main component of a refrigerator that needs power is the compressor. It is essentially a pump which is driven by a motor that 'sucks in' the refrigerant from the evaporator and compresses it in a cylinder to make a hot, high-pressure gas. The ***hum*** you hear when the fridge is on is that of the compressor working.

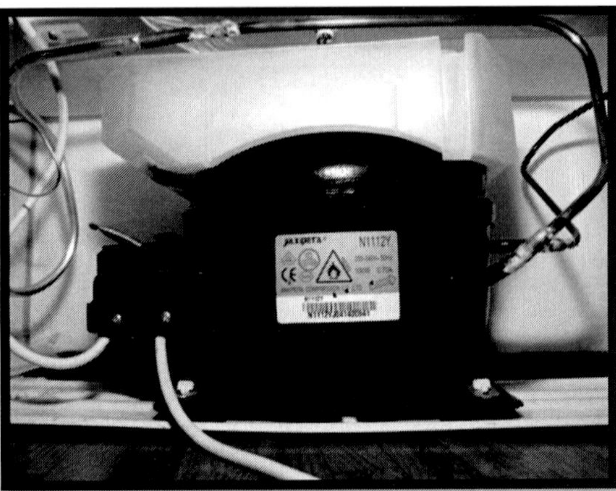

- **Evaporator:** This part cools the material stored in a refrigerator. It consists of finned tubes (made of metals with high thermal conductivity to maximize heat transfer) that absorb heat blown through a coil by a fan. The evaporator absorbs heat from the stuff kept inside, and as a result of this heat, the liquid refrigerant turns into vapor.

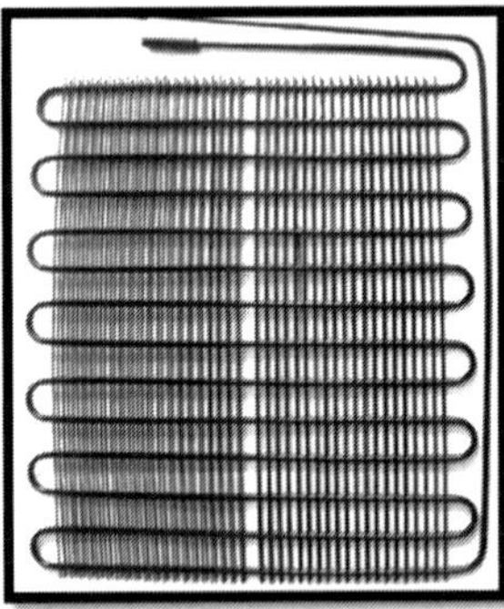

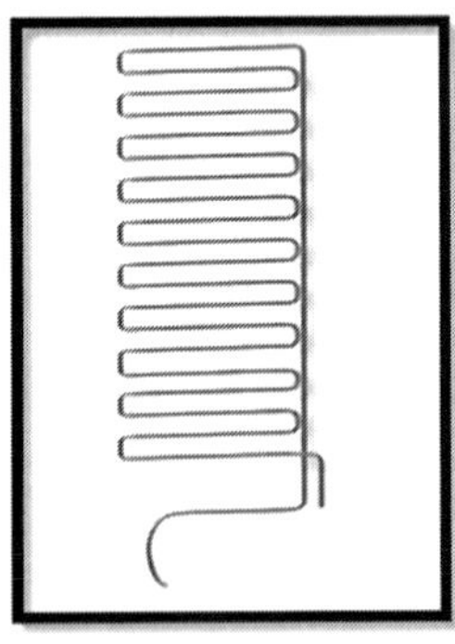

Evaporator

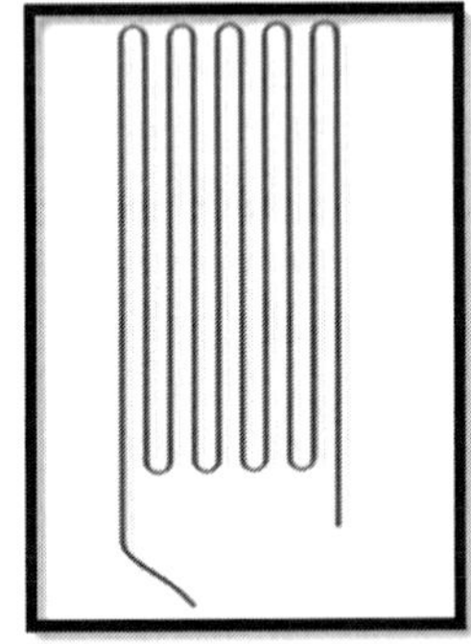

Condenser coils

Condenser: The condenser consists of a coiled set of tubes with external fins and is located at the rear of the refrigerator. It helps in the liquefaction of the gaseous refrigerant by absorbing its heat and subsequently expelling it to the surroundings. As the heat of the refrigerant is removed, its temperature drops to condensation temperature, and it changes its state from vapor to liquid.

- **Refrigerants:** Also referred to as coolant, it is the liquid that keeps the refrigeration cycle going. In fact, it is a specially developed chemical that is capable of alternating between being a hot gas and a cool liquid. In the 20th century, fluorocarbons, especially CFCs, were a common choice as refrigerants. However, they are being replaced by more environment-friendly refrigerants such as ammonia, R-290, R-600A, etc.
- **Temperature controlling devices:** These types of devices are used to control the temperature in the refrigerator. A thermostat is attached to the evaporator which helps to maintain a specific temperature. The thermostat controls the temperature of the fridge by switching on-and-off the compressor.

Refrigerator Function: How Does A Refrigerator Work?

Here is how the components of the refrigeration cycle interact to cool your food:

- The compressor constricts the refrigerant vapor, raising its pressure and temperature, and pushes it into the coils of the condenser on the outside of the refrigerator.
- When the hot gas in the coils of the condenser meets the cooler air temperature of the kitchen, it becomes a liquid.
- Now in liquid form at high pressure, the refrigerant cools down as it flows through the expansion valve into the evaporator coils inside the freezer and the fridge.
- The refrigerant absorbs the heat inside the fridge when it flows through the evaporator coils, cooling down the air inside the fridge.

- Last, the refrigerant evaporates to a gas due to raised temperature, and then flows back to the compressor, where the cycle starts all over again.

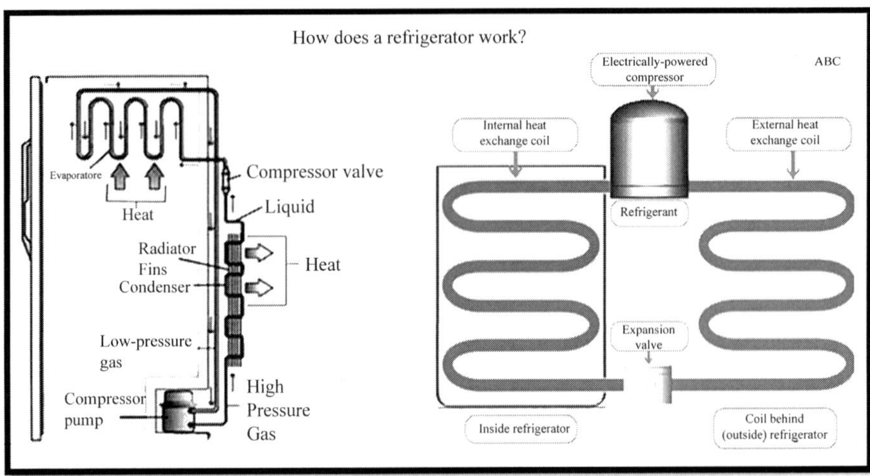

The compressor, which is a critical component of the refrigerator, compresses the refrigerant gas. As it undergoes high pressure, the gas heats up. Now, this gas transports to the condenser coils (thin radiator pipes) located at the back of the fridge, where the coils help dissipate its heat so that it becomes cool enough to condense and convert back into its liquid phase. Because the heat collected from the food items is given off to the surroundings via the condenser, it feels hot to the touch. The high pressure liquid that we have now flows through the expansion valve. Think of the expansion valve as a small hole. On one side of the hole is high-pressure refrigerant liquid. On the other side of the hole is a low-pressure area (because the compressor is sucking gas out of that side). After passing through the expansion valve, the liquid experiences a drop in the pressure. As a result, it becomes a cool (to the tune of -25°F) gas. When this cool gas flows through the expansion coils (installed inside the fridge), it absorbs heat and therefore make the insides of the fridge cold. This low pressure refrigerant gas is once again sucked by the compressor, and the whole cycle repeats, keeping the contents of a refrigerator perpetually cold.

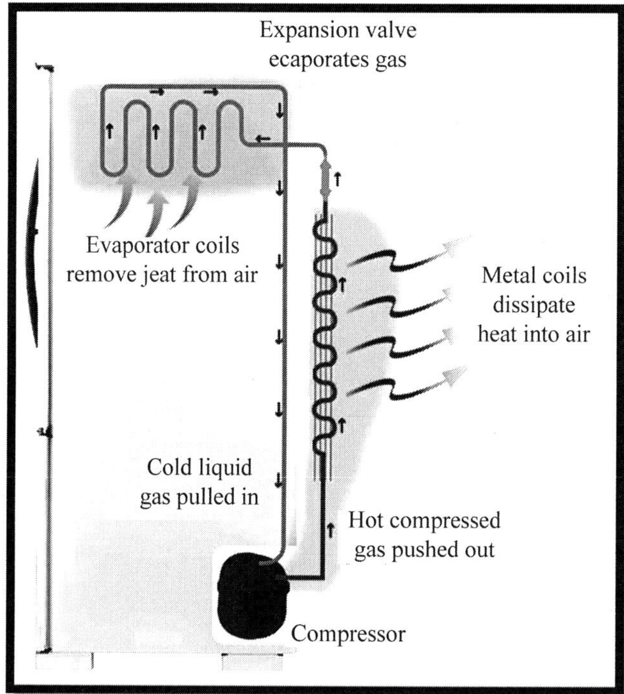

Advantages

Refrigerators are one of the most widely used home appliances over the years. You would hardly find any household without a refrigerator. Earlier, refrigerators didn't have many variations, but with the developments in technology, they have been modernized to a great extent. There are many different types of refrigerators available in the market now. All of them differ from each other in terms of price, features, shape, size, and capacity. Consumers can very conveniently choose the right refrigerator for themselves.

- **Food preservation:** They help in keeping your food and other edibles safe so that they can be put to use some other day. They prevent the food from getting spoiled and wasted. Refrigerators are not just to let you preserve cooked food, but also raw vegetables, fruits, spices, and so many other things.

- **Cooling:** Refrigerators provide cooling, which is necessary to keep your food fresh and safe. In the summer season, refrigerators also keep your water and other drinks cool so that you can enjoy drinking them to the best. They also have freezers which let you freeze water into ice, which can be used for many other purposes.

Safety Tips

Refrigerator is an electrical appliance and a pretty expensive one. So it is always better to be careful. You will need to take some precautions.

During installation

- **Dedicated Socket:** Never use extension cord to supply power to your refrigerator. Your refrigerator will run 24 x 7, for 365 days. So it is better to get a dedicated socket. Also, the extension cord will have multiple sockets. You might sometimes use those sockets. If you do, you will increase the risk of causing a fire. Because using multiple devices from one power outlet increases the chances of fire. So, get a dedicated socket for supplying power to your fridge.
- **Power plug position**: Be careful while positioning your power plug. If you are not careful, chances of your fridge catching fire will increase.
    - *Do not allow the power plug to face downwards.*
    - *Do not allow the power plug to get squeezed at the back of the fridge.*
- **Power plug condition**: Be careful while inserting the plug into the socket. You will see sometimes, some water and dust accumulate over the pins. You must clean them. Wipe them with a clean dry cloth. Once you clean them to fit it correctly and tightly into the socket.
- **Power supply cord:** Now that the power plug is inside the socket, it is time for the cord. A damaged cord will lead to an accidental fire. You can also get some electric shock if you touch the cord if it is damaged.
    - Power supply cord should not be bending.
    - Power cord should not be pressed under heavy objects.
    - Power cord length should not be extended or modified.
- These 3 things regarding the power supply cord should be kept in mind. Now some of them you have to check in a periodic manner like say after 3 months, it is good practice to check the cord in a periodic manner. But always remember to switch off while checking the cord.
- **Condition of power supply**: Do check the condition of both the power plug and the cord. You will need to do it in a periodic manner. If at any given point you see, power supply cord and the plug has damaged, or has holes or is loosed. Then the time has come to call the service center. Do not overlook it. It can have some very serious implications.

- **The place for the refrigerator**: Choose a place where you will be sure, that the fridge is not exposed to water. Make sure that there are no windows nearby. There should not be any wash basin or pipe near the refrigerator. Install fridge away from heat-producing kitchen appliances and equipments such as microwave or gas range.

During Usage

- **What not to place above the refrigerator:** The refrigerator itself takes up lots of space. We also will have access to space above the refrigerator. But use it wisely.
  - You should definitely avoid keeping some stuff above the refrigerator.
  - Avoid keeping Heavy dangerous objects above the refrigerator.
  - Avoid keeping electrical appliances above the refrigerator.
  - Do not keep harmful chemicals above the refrigerator.
  - Do not keep flammable substances above the refrigerator.
- **What not to put inside the refrigerator:** Do not put or store the following substances in your refrigerator.
  - Alcohol
  - Benzene
  - Ether
  - LPG Gas
  - Medicinal sprays.
  - Sprays containing alcohol (ethanol, methanol, ethyl glycol etc.)
  - Inflammable substances
  - Glass bottles in a freezer
- Use fridge for the right purposes. Avoid putting the above mentioned things inside your fridge and be safe.
- **How not to clean the fridge:** Cleaning the fridge is mandatory. Keeping it clean and properly maintaining it will increase the longevity of the appliance. But while doing so we need to be careful and avoid certain substances.
  - Avoid cleaning the fridge with water. Especially at the back.

  - Do not clean the fridge with benzene or a thinner.
  - If you cover the fridge, then do not cover the exit hole of cold air.
- But be sure to clean your fridge. Cleaning it will help you get rid of any bad smells, which is a common occurrence in a not well maintained fridge.
- **What not to do near the fridge**: At least keep a distance of **30 cms** from the fridge to any other substance. If it is a heating device, then more the distance, better it will be.
  - You should not use combustible sprays near the fridge.
  - You should not put or use heating devices near the fridge.
  - If you want to put a microwave oven or stove near it, keep some distance between them.
- **What not to do**: These are a list of things you should not do. This will prevent any accident from happening.
  - Don't give children easy access to the fridge. Especially the naughty ones.
  - Do not touch the power supply cord or plug with wet hands.
  - Do not touch food containers inside the fridge with wet hands.
  - Do not put your hands below/ bottom of the fridge.
- **When to call the service center:** If you find any fault in your fridge, call the service center. In some cases, you just cannot avoid it. Some of those instances are.
  - Power plug and supply cord damaged.
  - In case, of gas leakage, you should open the windows to ventilate the air and immediately contact the service center and say urgently.
  - In case of a strange smell, disconnect the fridge and immediately call the service center.
- **What to do**: Go through the **user manual** thoroughly before using the fridge. Place it safely on the selected position and use it safely.
- Also if you are leaving for a long vacation, you should switch off and unplug the refrigerator. This will prevent any unwanted incident from happening.

Exercise-1

Prepare two dishes and keep it in different brands of refrigerator for testing. Explain the process of preparing the dishes and compare the dishes on the basis of 9-point Hedonic Scale.

1-Like Extremely

2-Like Very Much

3-Like Moderately

4-Like Slightly

5-Neither Like nor Dislike

6-Dislike Slightly

7-Dislike Moderately

8-Dislike Very Much

9-Dislike Extremely

Recipe

Testing of Prepared Dishes on Hedonic Scale

S.N.	Parameters for evaluating the prepared dish				
	Appearance	Texture	Aroma	Color	Taste

Results

- **Best Appearance:**
- **Best Texture:**
- **Best Aroma:**
- **Best Color:**
- **Best Taste:**

Conclusion: ____

____

____

____

____

____

____

____

Remarks: ____

____

____

____

____

____

____

____

Signature

20

Washing Machine

Objective: To study the construction, functioning and working principle of washing machine.

Introduction to Washing Machine

A washing machine laundry machine, clothes washer, or washer) is a device used to wash laundry. The term is mostly applied to machines that use water as opposed to dry cleaning (which uses alternative cleaning fluids, and is performed by specialist businesses) or ultrasonic cleaners. The user adds laundry detergent which is sold in liquid or powder form to the wash water.

Construction and Working Principle

Two of the main types of washing machine are the fully automatic and the semi automatic machine. Fully automatic machines are more convenient, but semi automatic machines allow you more control (much like you have with hand washing), and they also tend to be a little more budget-friendly. Fully automatic machines change from one cycle to another, taking the clothes right through from start to finish without any additional input from you.

The working of the washing machine is quite high tech, and includes a set sequence of washing, rinsing, and spinning (plus a pre-wash if desired). During the wash cycle, the machine releases the detergent and fills the drum with water while moving the clothes around to make sure each item is thoroughly cleaned. This soapy water is then drained, and clean water is added for the rinse cycle, which removes the remaining soap. The water is drained once more and the machine spins quickly which removes around 80 percent of the excess water in the clothing.

Each cycle clicks over to the next automatically, making the fully automatic washing machine one of the most advanced washing machines on the market. The working principle of a semi automatic washing machine is quite different. While the cycles are quite similar, the way the machine works contrasts with the workings of the fully automatic machine. These machines work due to manual input – you tell the machine exactly what to do. This does have its advantages, such as allowing you to use only the water you need, and it gives you more control over how your clothes are washed. Once you've added your detergent and water, you use the settings to tell the machine to agitate, which acts as the semi automatic washing machine's version of the wash cycle. Clothes can then be moved to the second tub for the spin cycle. Both cycles use an agitator that has small blades for moving your clothes around in multiple directions.

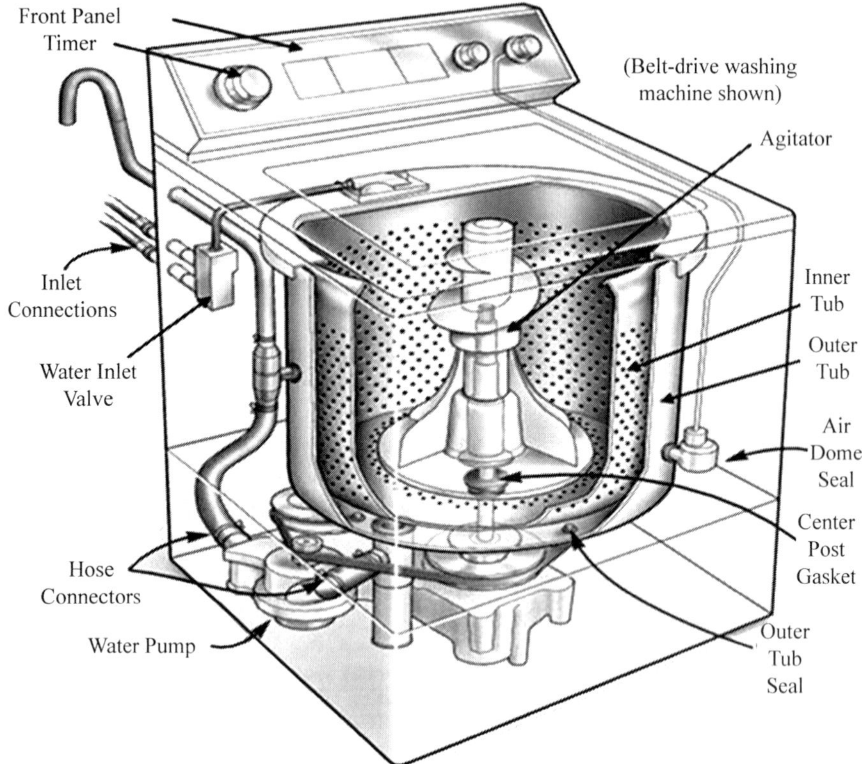

Internal Parts of Washing Machine

- **Parts of A Washing Machine Water pump.** This circulates the water through the machine, rotating in two directions. It's used for circulating the water through the wash cycle and also for draining the water during the spin cycle. Water inlet control valve. This is located near the water inlet point, which is opens and closes automatically when you load the clothes, depending on how much water is required.
- **Drum**. Did you know that washing machines actually have two tubs? The one you see where the clothes are loaded is the inner drum, which moves around the washing machine and is perforated with holes to

allow the water in and out. The outer tub contains the inner drum and the water, stopping it from leaking into the rest of the machine and supports the inner drum.

- **Agitator or paddles**. This is located inside the tub of the washing machine and helps perform the cleaning of the clothes. Most fully-automatic washing machines have these paddles on the rotating inner drum which is controlled by a rotating disc, whereas semi-automatic washing machines use an agitator that rotates within the machine to produce a current in the machine. Either way, these are designed to move the clothes around during the wash to allow the detergent to work or remove dirt particles and soiling from your clothes, helping the clothes rub together while washing.
- **Washing machine motor**. This is combined with the agitator or the disc that turns the drum, it produces a rotator motion. This is basically the mechanism that gets your machine going.
- **Drain pipe**. All the dirty water from your washing is expelled from the machine via the drain pipe.
- **Printed circuit board (PCB)**. This is where you'll find main the electronics that operate the machine from electrical components to circuits. These can be programmed and help operate the machine, acting as the artificial intelligence for the washing machine, sometimes even deciding on the time needed for rinsing or washing.
- **Timer.** This helps set the wash time for your clothes, which can be set manually or automatically.
- **Heating element**. This heats the water up in the washing machine to the desired temperature.

Different Types of Washing Machines

- **Fully-automatic washing machines:** A fully-automatic washing machine does everything at the touch of a button. This washing machine comes with a single tub that acts as a washer, dryer and also rinses the clothes. These can come as front and top load machines, and the programs and efficiency can also vary between the machines, depending on the brand.
- **Semi-automatic washing machines:** Unlike the fully-automatic washing machines, semi-automatic washing machines are only partially automatic washing machines. This means that you will need to do

some parts manually, like fill up the water in the tub before starting the machine, or if you have a twin-tub semi-automatic washing machine, you'll need to move the washed clothes into the drying tub. These are always going to be top loading.

Fully-automatic

Semi-automatic

- **Front load washing machines:** A front load washing machine is a fully-automatic washing machine that loads from the front side. These machines usually consume less water and energy, and usually yield better laundry results than their top load counterparts.

Top load

Front load

- **Top load washing machines:** Both manual washing machines and fully-automatic washing machines come in a top load form. These kinds of machines are loaded from the top. This is good for people who want a machine in which they don't need to bend down completely to load the clothes in the machine. Fully-automatic top load washing machines usually take up less space, so they're good for people looking to fit a washing machine into a narrower space.

Washing Procedure

1. Press the power switch ON (Lamps on the panel will light up.)
2. Set the water level selector.
3. Put the detergent
4. Put the clothes in the tub and close the lid.
5. Select a program by pressing the program select switch. The duration of wash, rinse and spin is automatically selected.
6. Now press the START/HOLD switch. When the operation starts, the program lamps will start blinking.
7. When operation ends, buzzer will sound six times intermittently. At the end of the operation, the normal lamp will still remain ON. Incase if the machine is not switched OFF within 10 minutes of completion of the operation. It will turn OFF automatically.

Precautions on Installations

- Always install the machine on a rigid horizontal floor.
- In case of wobbling, adjust leveling knob to ensure smooth Wobble-Free installation. When the floor is inclined, use (black) leveling pads.
- It is hazardous to install the machine in a damp environment such as a bathroom or such places where it can be exposed to rain water and direct sunlight.
- Insert the feed water hose to water inlet part and then tighten it. Connect the hose obliquely for smooth flow of water.
- Connect the drain hose to the drain outlet and clamp it with the hose band. If required an additional drain hose can be attached. When extending the external drain hose, do not increase its length to more than 3m, or its height to more than 20 cm.

Precautions While using Washing Machine

- Connect the power cord to electrical outlet where there are no other loads.

- Switch OFF the power and then remove the plug.
- Protect the power cord from damages caused by heavy objects.
- Do not put your hand in the rotating tub.
- When the lid is opened while spinning. The tub will automatically stop within 15 seconds.
- Avoid washing clothes which are cleaned or cleaned or soaked with flammable or explosive substances, in the washing machine
- Avoid splashes of water on the electrical switches.
- Never allow children, to fiddle or play with the washing machine.
- Make it a point to remove pins, hair pins, coins etc. from clothes prior to putting them in the tub, as the may damage the machine & clothes.
- Avoid putting hot or burning objects on or near the machine.
- Hot water is not recommended for washing, as it results in the weakening and the clothes life is reduced. However, lukewarm water may be used. To get optimum results use a good detergent.

Exercise-1

Identify and name the following equipments:

Conclusion: ______________________________

____

Remarks: ______________________________

____

Signature

21

Dishwasher

Objective: To study the construction, functioning and working principle of Dishwashers.

Introduction

Washing dishes is not the most rewarding task. Cooking can be creative, but cleaning up afterward seems like a waste of time and leaves the person washing complaining about "dishpan hands." The development of the dishwasher has helped relieve some of the monotony, as well as the grease and grime.

Components and Functioning

It operates on a simple principle of washing dishes that have been placed on racks inside the machine with multiple jets of water. A dishwasher is a machine used to clean dishware, cookware and cutlery automatically. Unlike manual dishwashing, which relies heavily on physical scrubbing to remove soiling, the mechanical dishwasher cleans by spraying hot water, typically between 45 and 75 °C (110 and 170 °F), at the dishes, with lower temperatures used for delicate items. A mix of water and dishwasher detergent is pumped to one or more rotating sprayers, cleaning the dishes with the cleaning mixture. The mixture is recirculated to save water and energy. Often there is a pre-rinse, which may or may not include detergent, and the water is then drained. This is followed by the main wash with fresh water and detergent. Once the wash is finished, the water is drained, more hot water enters the tub by means of an electro-mechanical solenoid valve, and the rinse cycle(s) begin. After the rinse process finishes, the water is drained again and the dishes are dried using one of several drying methods. Typically a rinse-aid, a chemical to reduce surface tension of the water, is used to reduce water spots from hard water or other reasons.

In addition to domestic units, industrial dishwashers are available for use in commercial establishments such as hotels and restaurants, where many dishes must be cleaned. Washing is conducted with temperatures of 65–71 °C (149–160 °F) and sanitation is achieved by either the use of a booster heater that will provide an 82 °C (180 °F) "final rinse" temperature or through the use of a chemical sanitizer.

The modern dishwasher has features that cater to fine glassware or the toughest pots and pans; multiple cycles that clean, sanitize, and dry; and under-the-counter or stand-alone models for every size, use, and price range. It is far from perfect; tough foods may need personal attention before and after dishes and pans are cleaned in the dishwasher, and few owners of crystal glassware and fine china are willing to trust them to a machine. But the dishwasher, like other kitchen appliances invented and improved in the twentieth century, is a fixture in many kitchens of the twenty-first century.

The major components of a dishwasher are made of steel and plastic. The basic structure consists of a steel frame assembly and a steel door panel. Sheets of stainless steel are purchased and fabricated in the required pieces and shapes in the factory; both the door and the wrap-around cabinet for standalone models are purchased as coiled sheet steel that has been prefinished in several standard colours. Other small steel parts are designed in house but made by suppliers to the manufacturer's specifications. The racks that hold the dishes are also made of steel, but it is delivered to the factory as coiled wire. To coat the rack tines to prevent them from scratching dishes, the racks are dipped in plastic in the form of powder polyvinyl chloride (PVC) or nylon. The inner box that holds the racks and the washer arms is called the tub. It is a single piece (not counting the piece lining the inside of the door) that is injection-moulded in the plant. The injection moulding is done with pellets of calcium-reinforced poly-propylene plastic. This plastic is respected for its strength and for the fact that it is inert; that is, it won't react with chemicals like those in detergents and is resistant to water and heat. Many other parts including the basket for cutlery, containers for detergent, and the wash tower and spray arms are also injection moulded. Motors, pumps, and electrical controls and components are made by subcontractors in accordance with designs by the dishwasher manufacturer.

Functions of Dishwasher

Basically, a dishwasher is a that cleans and rinses dirty dishes. Humans have to load the dishes, add detergent, set the proper washing cycles and turn it on, but the dishwasher accomplishes a whole series of functions by itself. A dishwasher:

- Adds water
- Heats the water to the appropriate temperature
- Automatically opens the detergent dispenser at the right time
- Shoots the water through spray arms to get the dishes clean
- Drains the dirty water
- Sprays more water on the dishes to rinse them
- Drains itself again
- Heats the air to dry the dishes off, if the user has selected that setting

In addition, dishwashers monitor themselves to make sure everything is running properly. A timer (or a small computer) regulates the length of each cycle. A sensor detects the water and air temperature to prevent the dishwasher from overheating or damaging your dishes. Another sensor can tell if the water level gets too high and activates the draining function to keep the dishwasher from overflowing. Some dishwashers even have sensors that can detect the dirtiness of the water coming off the dishes. When the water is clear enough, the dishwasher knows the dishes are clean.

Although dishwashers are watertight, they don't actually fill with water. Just a small basin at the bottom fills up. There, heating elements heat the water up to as much as 155 degrees Fahrenheit(68 Celsius) while mixing in the detergent. Then a pump propels the water up to the spray arms, where it is forced out and sprayed against the dirty dishes.

Stages of Cleaning

There are three main stages to a dishwasher cleaning cycle:

1. **Pre-wash/rinse:** The first initial burst of warm water through the spray arm that gets all of the dishes wet, but isn't really aiming to do much

cleaning.Detergent can be added to the pre-wash part of a dishwasher cycle.Detergent can be added to the pre-wash part of a dishwasher cycle. Some dishwashers have a pre-wash detergent dispenser to add some additional cleaning power to this cycle, for when dishes are really, really dirty. The pre-wash is typically only a few minutes long.

2. **Main wash:** Just what it sounds like: the main part of the wash cycle. The water is heated, sprayed, collected, filtered, heated, sprayed, etc. Until the heating unit is turned off, while the spraying continues. At the end of the main wash, all of the water is drained.Depending on the cycle, the main wash can take anywhere from 20 to 60 minutes, and may repeat multiple times throughout the cycle duration.
3. **Final wash and rinse:** This part pulls in new, clean water and begins the heating/spraying/filtering/heating cycle anew. Just like the main wash, the heater is eventually shut off while the spraying continues.The final wash/rinse may or may not use detergent. Like the main wash, the final wash/rinse can take 20 to 60 minutes and may repeat multiple times throughout the cycle duration.

Types of Dishwasher

1. **Built-in Dishwashers:** Built-in dishwashers are the most popular and recognizable types of dishwashers. Built-in dishwashers are the most popular and recognizable types of dishwashers. A built-in model has an independent and permanent connection to your plumbing, unlike some dishwasher types that need to route water from sink plumbing. This dishwasher slides into place and is secured in a spot under cabinets, which help buffer noise during wash cycles. The sides of a built-in dishwasher are unfinished, so we can remove it easily after disconnecting the plumbing and unscrewing two connection points at the front of the dishwasher. Standard built-in dishwashers measure 24-inches wide. Due to their popularity, 24-inch built-in dishwashers are the most widely available dishwasher option and have the broadest variety of features. A compact 18-inch built-in model is a popular choice for smaller homes and apartments. They are only made to go under benches and they typically do not come with sides or a top. There are many benefits of Built-in Dishwashers, including that they are very quiet due to the fact that they are surrounded by the cabinets in kitchen. They're also convenient to operate given how near they are to where food is prepared.

2. **Portable Dishwashers:** Portable dishwashers are not a common sight but solve a big problem (nobody likes doing dishes by hand) for people who live in small apartments, RV's and vacation homes. They're basically movable units on wheels, and they don't require separate plumbing to operate. Some smaller models can even be placed on a countertop. Simply hook it up to your kitchen faucet to operate. Portable dishwashers come in 24-inch and compact 18-inch variations.

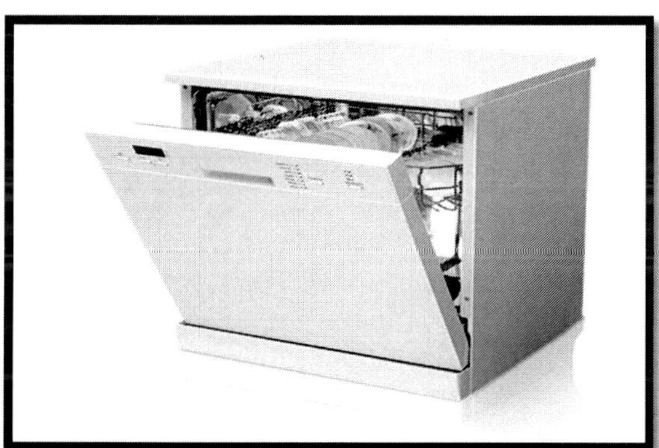

3. **Drawer Dishwashers:** A drawer dishwasher is the very latest design in dishwashers.Drawer dishwashers open like drawers rather than having a pull-down door, they're perfect for a home with a small amount of dishes because they're smaller than traditional models. This type ofDishwashers come in two formats, one drawer or double drawer dishwasher. Like a standard under-counter dishwasher, drawer dishwashers are about 24 inches wide and 24 inches deep. Single drawers are only between 16 and 17 inches high, while a double-drawer configuration is about 33 inches high. Each drawer holds about five place settings, and you can choose

single or double-drawer configurations.Both small and big kitchens can reap the benefits of the drawer dishwasher. In a small kitchen, a single drawer can be installed near the kitchen sink in a space normally reserved for a cabinet drawer.

4. **Countertop Dishwasher:** Countertop dishwashers are smaller than the portable dishwasher and are perfect for small family. Countertop dishwashers are easy to use and small enough to sit on your counter; they have a lower capacity than other types of dishwashers.It is a small dishwasher that sits on top of a counter or any other flat surface. It has a lot of the same capabilities as a full size dishwasher but is much more compact, portable, and uses less energy.Generally, these dishwashers use a quick connect adapter that connect to a standard sink faucet for water intake. They have a drain hose in the back of the unit that allows for drainage to run directly into a sink.The advantage of Countertop dishwashers is that the sizes vary and they don't need to be built in, which makes it easier to fit them on your free counter space.

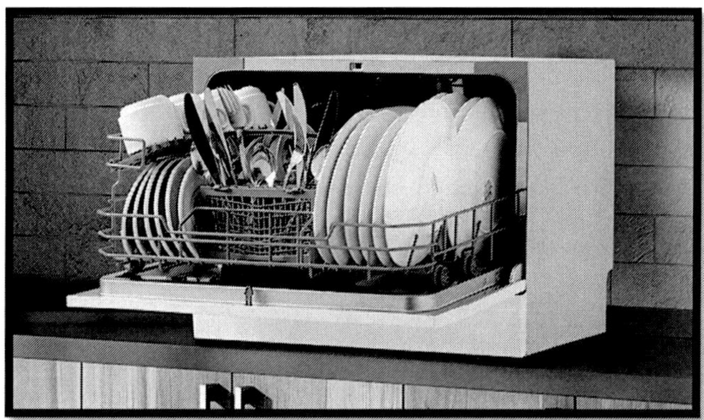

5. **Integrated Dishwashers:** An integrated dishwasher has a kitchen cabinet door attached to the front of the appliance. This makes it look like it's built into the kitchen furniture and 'hides' the appliance. Integrated dishwashers and integrated appliances in general, give a sleek appearance to any kitchen. They are often used in modern kitchens to help create that seamless and uninterrupted look. There are two types of integrated dishwashers: **fully** or **semi-integrated**. A **fully integrated dishwasher** has a full-sized kitchen cabinet door attached to the front of the appliance, completely hiding the machine. The control panel is accessed at the top of the dishwasher once opened, forming the top edge of the door. A **semi-integrated dishwasher** has a slightly shorter kitchen cabinet door attached to the front, leaving a strip exposed at the top. This is where the control panel and appliance handle are located. You do not have to have the door open to operate a semi-integrated dishwasher. Both fully integrated and semi-integrated dishwashers are available in a standard 600mm width and a slimline 450mm width.

Size of Dishwasher

- **Full size:** Full size dishwashers have a large capacity, They come in 2 options; the smallest version can usually hold up to 12 place settings, whereas the largest models holding up to 16.In full size dishwasher, you will save water and energy as you can wash many items at once.

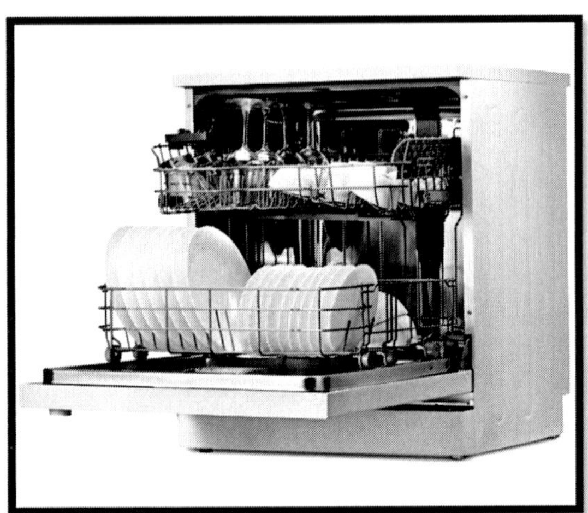

- **Slimline Dishwashers:** Slimline models are ideal for small or narrow kitchen. Slim line dishwashers are similar in appearance to the Freestanding Dishwasher however they are less in width which gives it slim line look.Slimline dishwashers offer similar cleaning power as their large counterparts but from slimmer design.

Exercise-1

Answer the following questions:

1. How do dishwashers wash the dishes?

____

____

____

____

____

____

____

____

____

2. List down the main components of a dishwasher.

Exercise-2

To understand the use of different parts and functioning and to check the performance of a dishwasher, wash the dishes in it and evaluate the dishwasher on the following parameters and give rating to each parameter on 9-point hedonic scale.

Rating	Parameters for evaluating				
	Cleaning	Washing time	Easy to use	Safety	Functioning
1-Like Extremely					
2-Like Very Much					
3-Like Moderately					
4-Like Slightly					
5-Neither Like nor Dislike					
6-Dislike Slightly					
7-Dislike Moderately					
8-Dislike Very Much					
9-Dislike Extremely					

Conclusion:

Remarks:

Signature

22

Electric Cooking Range

Objective: To study the parts, functioning and performance of electric cooking range.

Introduction

Electric Cooking Range is also known as an electric stove or electric range; is a stove with a built-in cooktop and with an integrated electrical heating device to cook and bake. Electric stoves became popular as replacements for solid-fuel (wood or coal) stoves which required more labor to operate and maintain. "Cookstoves" (also called "cooking stoves" or "wood stoves") are heated by burning wood or charcoal; "gas stoves" are heated by gas; and "electric stoves" by electricity. Unlike gas stoves, electric stove tops don't produce a flame. Instead, an electric current causes a heating element to activate. If your stove has coils, the heat transfers directly from these coils to your cookware. If your stove has a glass or ceramic top, the heat emanates to this cover and then transfers to your pots and skillets.

Parts and Functioning

Electric stoves generally have a thermostat, spiral/coil-shaped burners on top and heat-producing elements for baking and broiling inside. The standard electric stove has two large burners and two small burners, which are connected to the main part of the stove where the electricity is controlled by the dials on the front of the stove. Inside the oven, the bottom heating element is for baking and is controlled by the main thermostat on the front of the oven, which can be digital or analog. The top heating element for broiling does not have a refined thermostat control, so when it is turned on, it heats up swiftly. This is why ovens should never be left to broil unattended.

Electric stoves convert electricity to heat and serve as a replacement for gas stoves. When a burner dial is turned on, the electricity runs through the coil and the metal rapidly heats up. The dials control how much electricity runs through the coils, and thus how hot the burners get. The heating elements inside the oven function much the same way, with electricity making them hotter and the thermostat controlling how much electricity is fed to the baking element. Broil elements generally only have two settings: off, and broil.

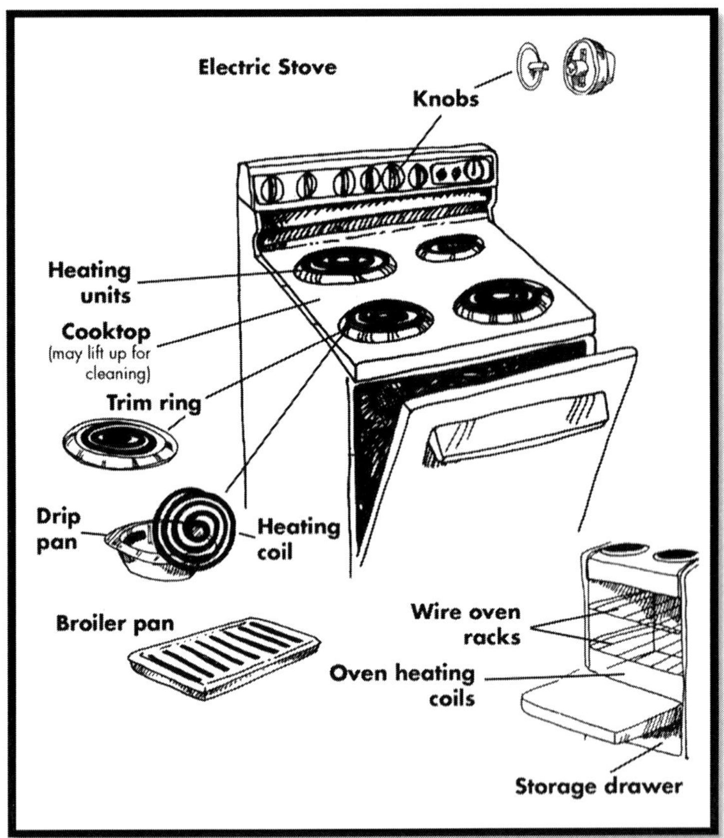

- **The Electric Stove Top Circuit:** Each heating element on your electric stove (typically four or five in total) connects to its own control switch. When this switch is turned on, it closes the circuit by connecting two "legs," each with 120 volts of alternating current. Electric current travels down each leg, and when this current collides in the heating element, heat is produced. A temperature switch also exists as part of this circuit, which allows you to easily adjust the heat output for your electric stove top. When the specified temperature is reached, the circuit will open and current will no longer be supplied to the heating element. But, when the temperature cools down too much, the circuit will close again to supply more heat. Because each burner represents its own circuit, you can remove and replace one faulty burner instead of replacing the entire stove.

- **Conventional Coils vs. Radiant Heat:** All electric stoves use the same basic circuitry to produce heat. But, in the case of ceramic stove tops, the heating element is hidden under a smooth surface. In terms of performance, the major difference is that conventional coils may heat up their surroundings, whereas a smooth-top stove is more efficient because it tends to heat just the pot. Other reasons why people may decide to choose radiant heating versus an electric coil stove include ease of cleaning and aesthetics. However, conventional coils are much easier to access in order to replace or repair them, and they tend to be cheaper than ceramic or glass stove tops.

Advantages

Some people prefer electric stoves because they are easier to use, don't require turning a dial to light the gas, nor are people in danger of accidentally turning off the heat while adjusting it. Electric stoves have a few advantages over gas stoves.

- **The Electric Stove Has An Elegant and Contemporary Shape:** The first advantage that you can see from an electric stove is that it has an elegant and contemporary shape that is very different from a gas stove. In fact, the electric stove has a slimmer appearance that makes it look modern.
- **Safer to Use Inside Confined Space:** Many people switch to using electric stoves because they feel safer when used in a closed room. You know, gas stoves are at risk of gas leaks that can lead to explosions and fires. While on an electric stove, this terrible thing is nothing to worry about because it uses electricity instead of gas. Not only that, electric stoves are considered safer because they don't feel hot when touched directly by hand.
- **Equipped with Many Modern Features:** Not only is it modern in shape, but electric stoves also have a variety of advanced features. The goal is to adjust the temperature produced so that cooking activities will be more effective. The features and components of the electric stove that you can enjoy include timers, automatic pan size, safety shut off, to boost.
- **Easy to Clean:** It is undeniable that the gas stove always leaves burnt marks around the installation area, so you have to spend more time so that the stove looks as clean as before. This sort of thing will not happen when you are using an electric stove. Because almost all brands

of electric stoves use glass material with a flat surface. Thus, you only need to wipe the surface until it is completely clean.

- **Electric Stoves Produce Heat More Evenly:** Who would have thought, the heat generated by electric stoves is far more stable than gas stoves. This is what makes your cooking evenly cooked. Interestingly, the stable heat allows for much shorter and more effective cooking times. In fact, the bottom of the pan will not turn black from exposure to excessive heat.
- **Generated Heat Lasts Longer:** Almost all electric stove products have the ability to store heat for a while after being turned off. Even if the heat that remains isn't the same as when it was turned on, you can still use it to warm up small portions of food.

Disadvantages

Even though it is equipped with various advantages, it turns out that electric stoves also have disadvantages that you can consider before buying them.

- **The Price is Relatively Expensive:** When compared to a gas stove, of course, this one electric device is priced relatively more expensive. The reason is, the electric stove is equipped with a variety of sophisticated and modern features. In addition, the price of electric stoves is also influenced by the number of stoves available. The more stoves, the higher the price is set.
- **Requires A Lot of Electrical Energy:** As the name implies, this one stove uses electrical energy as a substitute for fuel. But unfortunately, the electricity needed is quite large, which is around 300 to 1,200 watts. So, you have to be prepared with bloated monthly electricity bills when you want to replace a gas stove with an electric stove.
- **Must Use Special Cookware:** Apparently, not all cooking utensils can be used on an electric stove. The stove surface that tends to come up certainly does not allow you to cook in a deep pan. Therefore, you must have all list of cookware with a flat bottom when you want to use the electric stove. Not only that, you can only use cooking utensils made of iron, steel, and stainless steel. This is because these three materials are able to provide a special interaction with the surface of the electric stove so that the cooking process is perfect.
- **They cook food slower than gas:** Direct flames from gas burners may heat your cookware faster than the heating elements in electric stoves, and it's easier to keep flames at a steady height and maintain a specific temperature.

Precautions While Using Electric Stove

The stove is a critical component of the household and life would be hard without it. We use it on a daily basis for small meals and large meals alike, but life can be dangerous if you are not practicing safety and caution while using a electric stove.

- Never leave cooking unattended.
- Keep it clean. Food particles and grease residue can ignite and cause a fire.
- Create a "Kid-Free Zone" that children are aware they are not allowed to enter.
- Before each use, do an oven safety check.
- Be sure you do not smell gas leaking from a gas stove. If you do, open a window and leave the room.
- Make sure there is nothing in the oven. An oven should never be used for storage.
- Make sure that there are no food particles or grease residue on the stove top.
- Position your oven racks before pre-heating to prevent scalding.
- Be sure to follow safety guidelines when using stove top burners.
- Use burners on the back as much as possible.
- Never use a pot that is too small or big for the burner.
- Never leave a burner on that does not have a filled pot or pan on it.
- Do not leave an empty pan on hot burners.
- Turn all pot and pan handles inwards. They can be pulled and knocked over when they are turned out.
- Close the oven door immediately after removing or testing food.
- Keep oven mitts and other cloth and paper out of the way of the burners.
- Use oven mitts and oven sticks to remove pots and pans from the stove top or oven.
- Keep a fire extinguisher, baking soda and metal lids nearby to put out fires quickly.

Exercise-1

Identify the name of different parts of the following equipment:

Exercise-2

Answer the following questions:

1. Is there a significant difference between using conventional coils versus radiant heat?

____

____

____

____

____

____

____

2. Are electric stoves better than gas?

Exercise-3

To understand the use of different parts and functioning and to check the performance of a electric cooking range, prepare a dish in it. Taste the prepared dish; evaluate the dish on the following parameters and give rating to each parameter on 9-point hedonic scale.

Rating	Parameters for evaluating the prepared dish				
	Appearance	Texture	Aroma	Color	Taste
1-Like Extremely 2-Like Very Much 3-Like Moderately 4-Like Slightly 5-Neither Like nor Dislike 6-Dislike Slightly 7-Dislike Moderately 8-Dislike Very Much 9-Dislike Extremely					

Results:

Conclusion:

Remarks:

Signature

23

Solar Cooker

Objective: To study the parts, functioning and working principle of solar cooker.

Introduction

A solar cooker or solar oven is a device which uses the energy of sunlight to heat food or to cook it or sterilize it by means of reflective panels that concentrate the light on to a dark coloured pot in an insulated box.

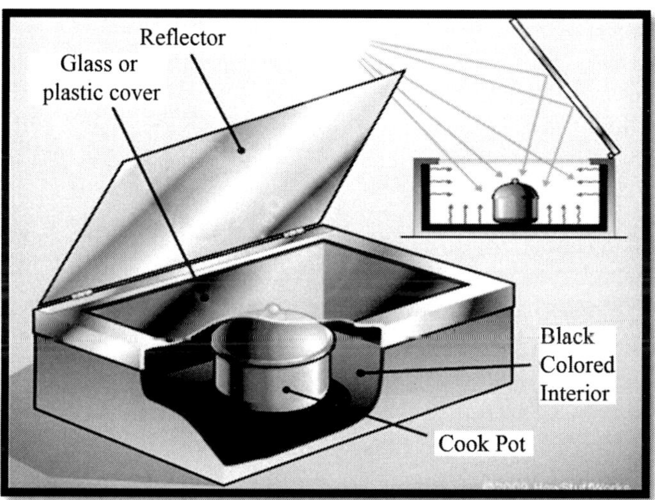

Parts and Functioning

Solar cooker consists of a square metallic box with a mirror fixed on the inner side of the lid. By keeping the lid open at a certain angle, this mirror is used to focus sun rays on the food kept in the round containers placed in the box. It cooks food with the help of solar energy and can save the conventional fuels to a significant amount. Solar cooker works on the principle that sunlight warms the pot, which is used for cooking the food. Now, this warming of the pot occurs by converting light energy to heat energy. Concave mirrors are used in these types of cookers because these mirrors reflect sunlight into a single focal point. Solar cooking is done by means of the suns UV rays. A solar cooker lets the UV light rays in and then converts them to longer infrared light rays that cannot escape. Infrared radiation has the right energy to make the water, fat and protein molecules in food vibrate vigorously and heat up. It is not the sun's heat that cooks the food, nor is it the outside ambient temperature, though this can somewhat affect the rate or time required to cook, but rather it is the sun's rays that are converted to heat energy that cook the food; and this heat energy is then retained by the pot and the food by the means of a covering or lid. The main advantage of the solar cooker is that it uses solar energy which is free of cost; it also keeps the food warm so that one does not have to heat it before eating.

There are three main components to most solar cookers:

- Concentration (reflection, or reflectance): Concentration of the sun's rays is performed most often by reflecting panels, petals and such surfaces that can "focus" or concentrate the rays of light (UV) to a point or concentration. These reflecting panels are usually made of materials that are shiny and reflective due to the substance used in their manufacture, such as silver, chromium and aluminum. Without a means to concentrate the sun's rays it can take longer to heat objects or surfaces, thus most solar cookers are constructed with reflector panels to speed up the process of heat accumulation.
- Absorption (ability to attract or hold heat): Absorption of the sun's energy (heat) in solar cooking is best achieved when a surface is dark in color, thus the most common solar oven interiors are usually black in color as well as the color of the cookware used for cooking the food. Dark colors absorb the heat, whereas light colors do not absorb heat well. And some colors can reflect the sun's energy away, such as silver for example. Generally, the best cookware for solar cooking is the dark, thin variety because it absorbs the heat (energy) well and because it is thin, it can then transfer the heat more quickly and evenly to the food.
- Retention (means or capacity to retain heat): Retention is the final principle in solar cooking. If a solar cooker is not well insulated and if it does not have a cover, or lid, then all of the concentrated heat (energy) and all of the absorbed heat would quickly dissipate into the air and be lost to the surrounding environment. A solar cooker must have the means to "trap" or hold the concentrated heat allowing it to accumulate and to "build up" to sufficiently high enough levels to be able to effectively cook.

Types of Solar Cookers

Basically, there are 3 types of solar cookers:

- **Box Cookers:** Solar box cookers (sometimes called solar ovens) are the most common and inexpensive type of solar cookers. These box cookers have a very simple construction and they are made of low cost materials. The outer box is often made of wood which consists of an insulated inner box, metallic cooking pots inside the box, double glass lid on the cooking tray, and a reflecting mirror fitted on the underside of the lid of the box. The heat is absorbed by the blackened surface and gets transferred to the food inside the pots to facilitate cooking. The solar box cooker, like other solar cookers, needs direct sunshine to operate and produces zero emission. However, the temperature is low and it cannot store and save solar heat for later use.

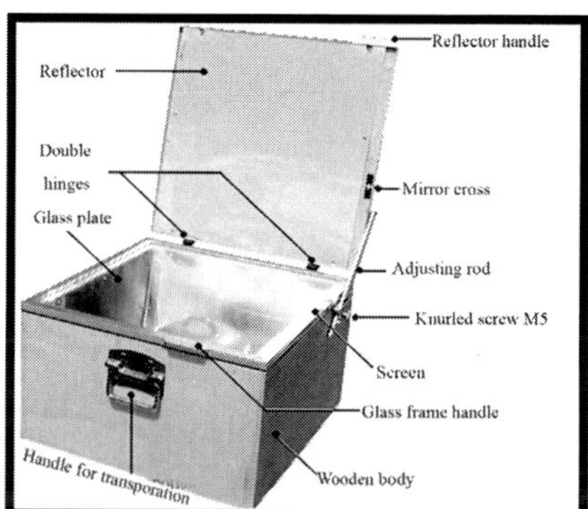

Panel Cookers: Panel Cookers have a flat pancl which reflects and focuses sunlight for cooking and heating. Panel cookers incorporate elements of box and curved concentrator cookers. The panel cooker is quite similar in operation to the solar box cooker. The same principles are employed but instead of an insulated box, panel cookers typically rely on a large reflective panel. Panel cookers are unable to collect and store sunlight for later use and they are unstable in high winds. Also it cannot retain much heat when the sun is hidden behind the clouds.

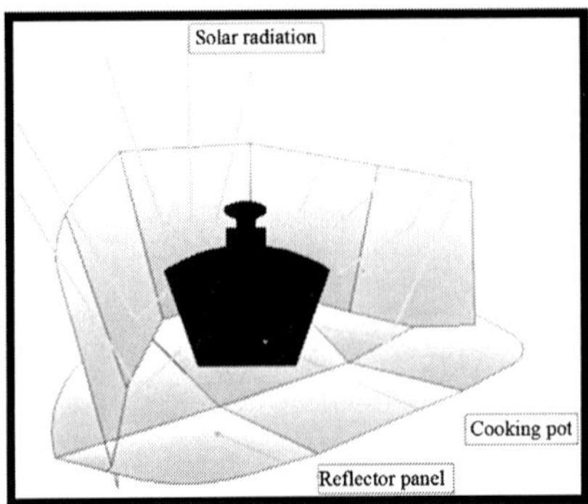

Parabolic Cookers: Parabolic Solar Cookers (sometimes called curved concentrator cookers) in comparison to solar box and panel cookers, they can reach much higher temperatures, 350 °C or higher, and can cook more quickly, but require frequent adjustment and supervision for safe operation. It needs more precision to focus the sunlight on the food in the pan. If the sunlight is not correctly focused on the food in the pan, the food will not be cooked well. Parabolic solar cookers use a parabolic-shaped reflector to direct sunlight to a small area in order to generate heat for cooking.

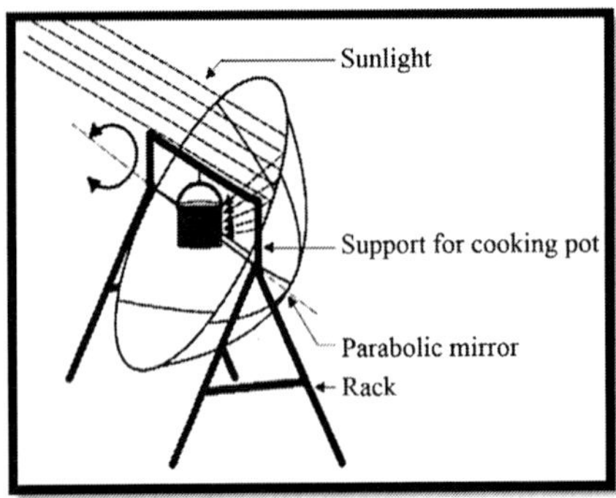

Precautions

- Do not look at the reflection of the sun at any time. It will damage your eyes.
- Once the stove is in alignment with the sun, and the pot is at the focal point, then there should be no light scatter.
- If there is 'stray' light then the stove may need adjustment such that all the reflected light hits the pot (this will improve your cooking as well).
- When finished, turn the cooker upside down so as not to dazzle anyone.

- Wear sunglasses whilst adjusting the cooking position.
- When adjusting the angle of the grill use thick gloves as the grill can get very hot.

Cooking Tips

- Black pots work a lot better than silver pots. The pot needs to absorb as much light as possible and silver tends to reflect the light. Dull or 'matt' finishes absorb more light than 'shiny' surfaces.
- Pots with close fitting lids keep the heat in and help the cooking process. Placing the stove in a sheltered area stops the wind from cooling the outside of the pot.
- Jars can be painted black to increase their efficiency.
- Thin metal heats up quickly. Granite ware cookware is popular.

Advantages

- High-performance parabolic solar cookers can attain temperatures above 290°C (550°F). They can be used to grill meats, stir-fry vegetables, make soup, bake bread, and boil water in minutes. Vacuum tube type cookers can heat up even in the clouds and freezing cold.
- Conventional solar box cookers attain temperatures up to 165°C (325°F). They can sterilize water or prepare most foods that can be made in a conventional oven or stove, including bread, vegetables and meat over a period of hours.
- Solar cookers use no fuel. This saves cost as well as reducing environmental damage caused by fuel use. Since 2.5 billion people cook on open fires using biomass fuels, solar cookers could have large economic and environmental benefits by reducing deforestation.
- When solar cookers are used outside, they do not contribute inside heat, potentially saving fuel costs for cooling as well. Any type of cooking may evaporate grease, oil, and other material into the air, hence there may be less cleanup.
- No health problem because there is no smoke.
- Reduces your carbon footprint by cooking without the use of carbon based fuels or grid electricity from traditional sources.

Exercise-1

Identify and name the following:

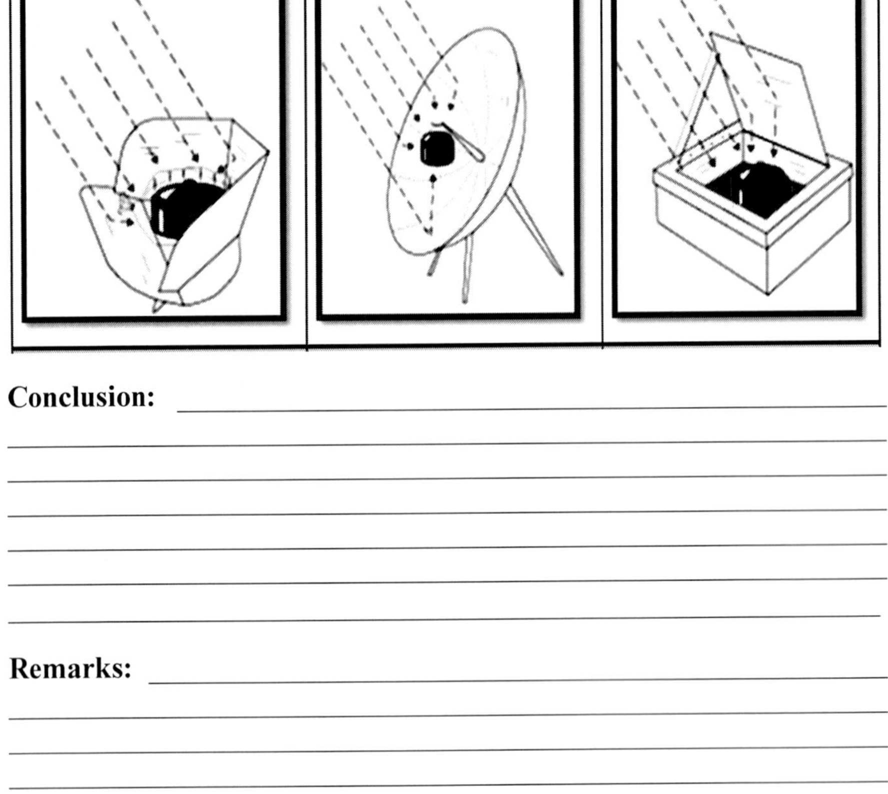

Conclusion: ______________________________

____
____
____
____
____
____

Remarks: ______________________________

____
____
____
____
____
____

Signature

24

Kitchen Layouts

Objective: To study about different types of kitchen layouts.

Introduction

To learn the design and arrangement of space in the kitchen which is the key in the pattern of living, the number of people to be served in the area, the size of the house and the equipment and furnishing in the kitchen. To know the most important aspect to consider in planning a kitchen i.e., the nature of the primary and secondary activities to be performed in the kitchen.

Kitchens

Kitchens come in a wide variety of sizes, shapes, and layouts, but well-designed kitchens have a few things in common. There are several crucial elements we should consider to help us design the most effective floor plan for our kitchen, whether we're planning a galley kitchen, U-shape kitchen, L-shape layout, island, or peninsula. In addition to planning the placement of appliances, we'll need to plan clearances, traffic flow, storage details, and even the swing of all doors. Space in kitchen may be divided into canters for specific activities. When planning a center, we should determine – 1) the activities 2) the processes in each activity 3) the supplies and equipment needed for the performance of process 4) the working heights and space – storage, work counter and floor space – necessary for the performance of the activity and 5) the relationship of the specific center to the other activity areas in the kitchen. Also, locate supplies and equipment to make the most efficient use of space, and locate the center in relation to other centers to fit into a logical flow of work. Attention in kitchen space has to be focused on four centers in the modern kitchen i.e., the sink, mix, refrigerator and range-serve centers.

Kitchen Size

An idea of total space, compared with work centers desired, may help us in working with our kitchen planning. Figures are generally used by kitchen planners:

- Small kitchen 8 feet by 10 feet to 10 by 10
- Medium kitchen 10 feet square to 10 x 12
- Large kitchen 10 feet x 12 feet and up

Arrangement

Centers can be arranged in any logical manner that considers the space, activities and the importance we attach to a particular center. Areas around the three basic pieces of kitchen equipment — the sink, range and refrigerator — are usually considered first. Then plan and locate these most important centers in the space we have to work with. We should try to place closely related centers together. Plan for routing our work from center to center. Think through a meal from planning, purchasing, storing, preparing, cooking, serving and eating, to cleaning up. Our kitchen plan should make these tasks flow easily from one work area to another without unnecessary walking or traffic interference.

- **Work triangle:** The work triangle concept can give us a basis to see how well your space planning will facilitate your work in the kitchen. The triangle formed by the three lines that connect the center fronts of the refrigerator, sink and range or cook-top as shown in the figure. We may disregard the separate oven, if we have one. The most desirable size is a total of 15 feet to 22 feet of the three sides of the work triangle. The absolute minimum is 12 feet and the triangle should not be more than 26 feet, according to kitchen planning research.

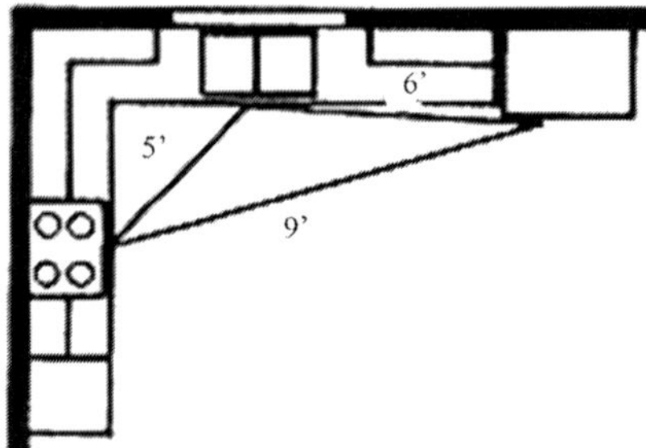

Other guidelines to use in checking the work triangle are:

- Plan major traffic routes outside the work triangle
- Distances between appliances should be no more than
    - 4 feet to 7 feet — refrigerator to sink
    - 4 feet to 6 feet — sink to range
    - 4 feet to 9 feet — range to refrigerator

- The work triangle is smallest with the Corridor (two-wall) and "U"-type kitchens. The corridor plan is generally a traffic way, while the "U"-type kitchen is dead-end space
- Clearance between base cabinets or appliances opposite each other should not be less than 48 inches. In larger homes (over 1400 square feet) this may go up to 60 inches or more
- **Sink center:** The sink center is the most frequently used area in the kitchen and should be central to other major centers you plan. In this center, we will have sink, water and drainage for food preparation and kitchen clean-up jobs. We may want to include a food waste disposer, an automatic dishwasher and water heater. For a right-handed person, 24 to 36 inches of counter space should be at the right side of the sink and 18 to 30 inches at the left side of the sink. If a dishwasher is placed to the left, 24 inches is adequate. Poor storage space is usually a problem in the sink center because of the plumbing lines and fittings, disposer and dishwasher. Some of the usual items to consider storing in this area are: tools to clean food (brushes, etc.), storage supplies for leftovers, dishwashing supplies and tools, pitchers, coffeemakers, cutting board and wastebasket. Cleaning supplies are stored in this area in many homes — if possible, we should try to locate these in another area or at least out of the reach of small children. The sink has to be durable, be able to withstand the aftermath of cooking adventures and serve as a stylish focal point at the same time. The sink center serves primarily for food preparation and doing activities that require a supply of water and drainage. The sink, for water in the performance of all types of tasks in the kitchen requires that the sink be easily accessible from all other work activities. Each task requiring supplies and utensils which are first used and water should be stored in this area. The sink center should be between the refrigerator and the range center or cooking area, for maximum efficiency. Since sinks get the most use and traffic of any spot in the kitchen, centrally located sinks and dish washers work best.Plan to have the most uncluttered countertop space in the sink center. The area in and around sinks is used for food-preparation tasks involving cleaning and cutting, as well as washing and cleaning up after meals. Place trash and recycling containers strategically so that tidy disposal happens effortlessly.Storage at sink centers includes places for everyday flatware, dishes and glassware, as well as activities involving waste management, dishtowel placement and storage, cleaning supplies, polishing and drying. Place cutting boards and knives at a midpoint between the sink and refrigerator so you can easily access items from

the refrigerator, cut and chop as necessary, and dispose of the remnants to the sink center, which holds the refuse containers.

- **Mix center:** The mixing center or food preparation center is ideally located between the refrigerator and sink center. If it is between the sink and range, it will involve more travel by the user. If possible, avoid a window in this area so we can have the needed wall storage. Most packaged, canned and bottled foods should be stored in this center as well as equipment, baking pans and tools used for food preparation. Length of counter recommended for mixing is a minimum of 36 inches. When it is possible, a lower counter in the mixing area makes a more comfortable working height for most homemakers. To see the height that would be best for our standing work at the mixing counter, measure 7 to 8 inches below your elbow height. To test this figure, improvise a work surface at this measured height and then work at this height long enough to decide if it seems right for us. Make needed adjustments and use this latter adjustment in your planning for the mix area.
- **Range-serve center:** Two configurations, a range or a cooktop with wall ovens, comprise the range center, where cooking food and preparation for serving takes place. Place these functions toward or near the dining room. Anything that involves the cooking process needs to be within the range center. Cooktops and ovens with ample countertop space next to or closely across from them provide a spot to set down hot items quickly. In immediate and obvious proximity to the range center, put potholders and other items that aid in handling hot cookware. Also use this location to transfer cooked food to serving dishes. Plan to store platters, bowls and other equipment used to get food to the table around the range center.
- **Refrigerator center:** This taller appliance is best located so it will not block movement from one work center to another, so is usually at the end of the work area. One common fault in this area is not providing counter space on the door opening side for convenience of putting in or removing things from the refrigerator. An 18-inch counter should be beside the door handle. The refrigerator location should not interfere with food transfer between counters and the dining table and near preparation centers if possible. It is helpful to locate this center near the entrance where you will bring groceries into the home. The refrigerator may be arranged in two ways in a kitchen. One as a separate center with counter and storage space, or in combination with another center. Some of the functions to be served by the refrigerator are refrigeration of perishable foods, storage of leftovers, preservation of beverages and

storage of frozen foods. Heat sources such as ovens, radiators and even direct sunlight can affect the energy efficiency of a refrigerator and potentially shorten its functional lifespan. When impacted by constant heat, a refrigerator will have to run more often to maintain appropriate internal temperatures. Fridges should never be placed directly next to a heat source or in locations affected by direct sunlight for long durations. The fridge should always be located in close proximity to a bench with ample available space. This will allow grocery loading to be prompt, meaning the door is open for a lesser amount of time, and when gathering ingredients for food preparation one can easily grab what is required and place down promptly on the bench. The shorter the time frame the door is open, the less impact on temperature variation within the fridge or freezer.

The Secret to Positioning Appliances in Kitchen Design

- **What do we need?**
  - We may love the idea of a massive refrigerator or a large, professional oven, but do you really need it is the question in designing kitchen. If we do not cook enough or have a large family to feed, these large appliances become nothing more than decorations that take up space. We should consider how our family and home function, how much space we have, and choose appliances to fit our kitchen design requirements.

- **Think about our work areas.**
  - The kitchen layout is typically based on work zones that allow particular activities to take place simultaneously, possibly with multiple people at once, without tripping over each other. This may include food preparation, clean up, dining and entertaining, and possible a study zone or pet area. The sink, main refrigerator, and stove should be in close proximity to each other. We require a minimum of 4 feet between each point to easily maneuver, and equally not more than 9 feet. This allows easy access to get food from the refrigerator to a preparation point, and then into the oven. A dishwasher is typically located near the sink. In a large kitchen design, you may have a second clean up sink and put the dishwasher in this area, perhaps in the kitchen island.

- **Consider the flow of traffic.**
  - Workflow is critical to smooth functioning of the kitchen. We need to have a clear path from one area to next when cooking or cleaning up,

without worrying about tripping over people or obstacles. Make sure you can easily move between refrigerator and food preparation areas, then to the oven and range, and finally to a clean-up area that includes the dishwasher. Consider putting a secondary beverage or snack refrigerator in a separate area so guests or kids can access food and drink without interfering with the cooking.

- **Leave enough space for doors and drawers.**
  - Allow ample space for opening and closing appliances, particularly the oven. We must have enough room for the oven door to open, and to safely maneuver around the oven door to get food out and to a landing point. Consider also the direction in which the refrigerator door swings, so we can easily put away groceries and access food when cooking. A refrigerator door must be able to swing wide enough so internal bins can be fully pulled out. Make sure an open dishwasher door does not block access to the sink or create a hazard.
- **Look at size and installation requirements.**
  - There are many things to consider when looking at the size of your chosen appliances. Do we have a large family or do we frequently entertain? Are we downsizing now that your kids have moved away? Buy appliances that meet your household's requirements and remember that bigger is not always better. Balance the size of appliances to other aspects of your kitchen, like kitchen cabinet space, countertop depth, and workspace. Appliances that are too deep stick out past cabinetry, which looks bad, interrupts traffic flow, and causes accidents.
- **Don't forget about small appliances.**
  - Many kitchens include coffee machines, slow cookers, toasters, stand mixers, and more. Make sure these small appliances have a dedicated home or easily accessible storage. Install coffee makers in a beverage station, give stand mixers a dedicated pop-up shelf, and store other appliances on pull out shelves. While planning any kitchen we should know the activities, size and other aspects of the family and most importantly we should plan with the proper and appropriate work-triangle so as to avoid the work traffic of the family around the kitchen.

Layout Plan for Different Shapes of Kitchen

Since the kitchen is one of the most active work areas of the home, it is important to select the right layout to complement your lifestyle and taste. There are five primary kitchen layout shapes:

1. One wall kitchen
2. Two wall/Corridor/Gallery kitchen
3. L-Shaped kitchen,
4. U-Shaped kitchen
5. Island kitchen

The ideal shape for the kitchen arrangement is 'U'. It consists of preparation and cooking centers on either side and the cleaning center in the middle. This is a compact arrangement and labour saving too. In the L-shaped kitchen arrangement of the work centers are placed on two adjoining walls. The one wall arrangement is suitable for small kitchen.Island layout is perfect if you plan to entertain.

L – Shape Kitchen

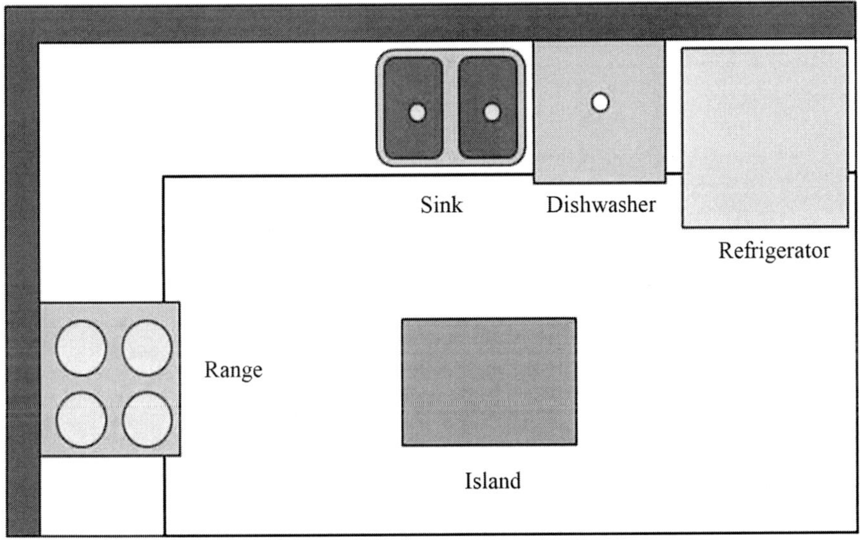

U – SHAPE KITCHEN

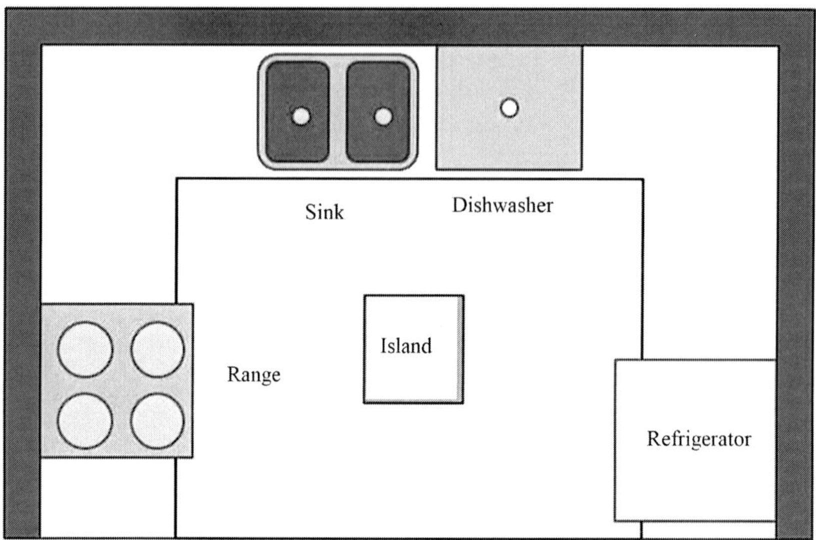

ONE-WALL KITCHEN

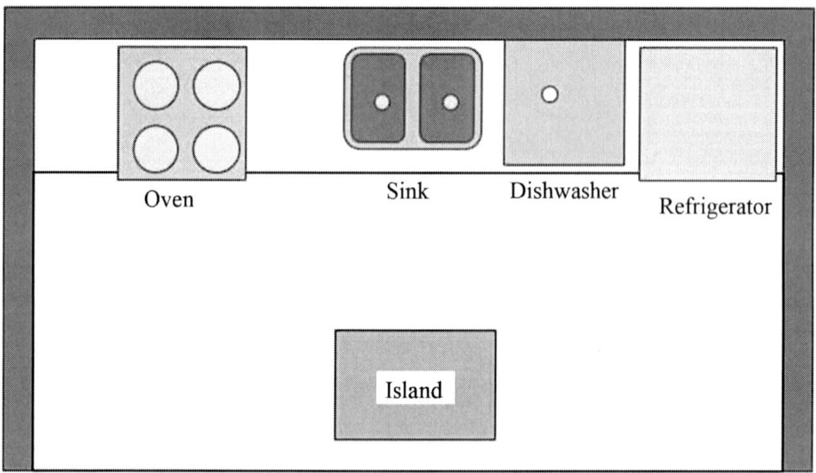

TWO WALL OR GALLEY KITCHEN

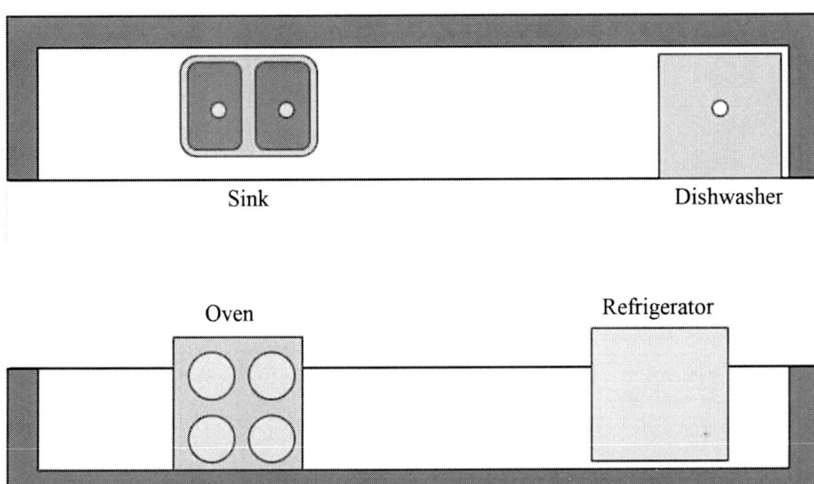

ISLAND KITCHEN

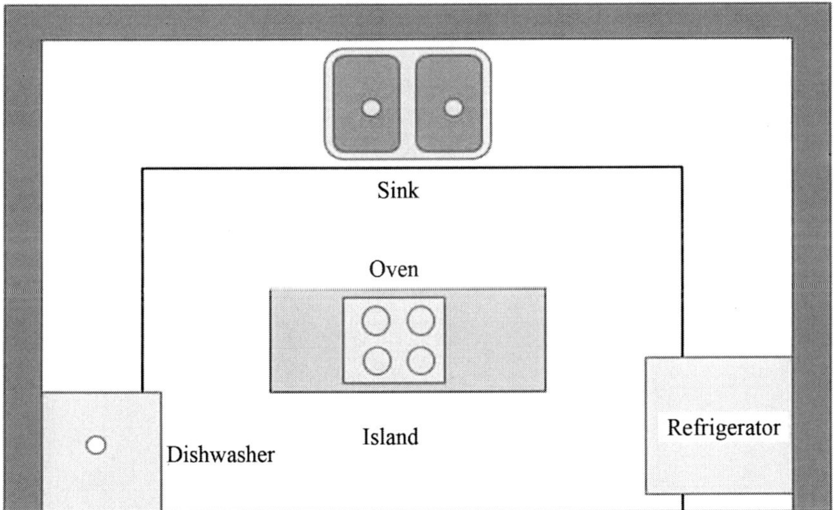

SHOWING WORK-TRIANGLE IN DIFFERENT TYPES OF KITCHENS

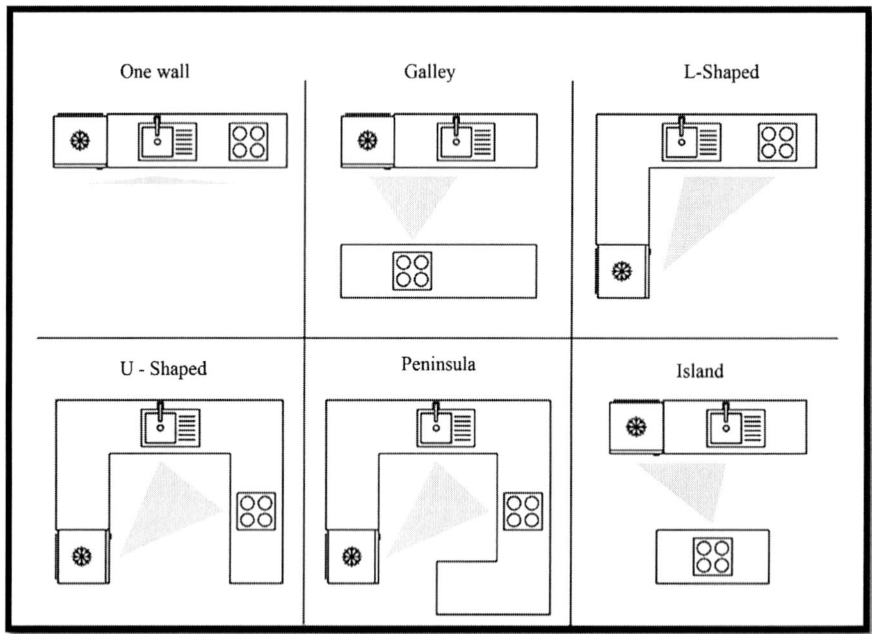

Exercise-1

Draw a plan layout of the kitchen with reference to the centres used in day today houses.

Conclusion: ____

____

____

____

____

Remarks: ____

____

____

____

____

____

Signature

25

Designing an Ergonomic Kitchen

Objective: Designing an ergonomic kitchen.

Introduction

Every homemaker spends a chunk of her time in the kitchen. Kitchen is the most functional part of the house. It needs to be carefully designed. Kitchen ergonomics comes into play to make the kitchen a comfortable place to work. Kitchen ergonomics is all about minimizing the movements and stress we have to go through while cooking. While designing your kitchen you need to keep in mind the heights of various surfaces, distances and placement of items to make your space functional, efficient and enable effortless cooking.

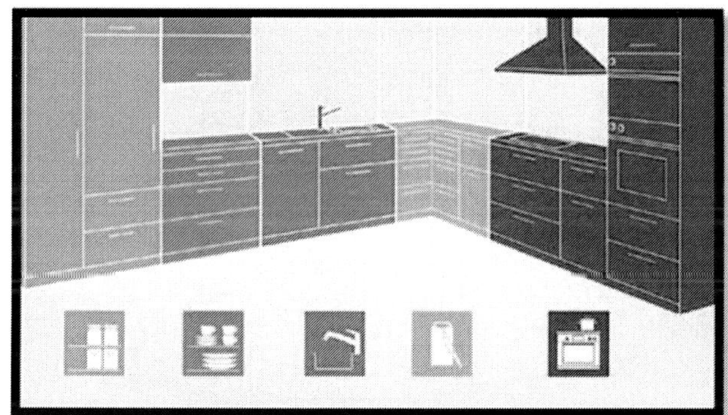

Kitchen is broadly classified into five different zones. The kitchen ergonomics revolve around these five zones. You can have a flexible work zone by organizing the different zones to work at ease.

1. Consumables food
2. Non-consumable food
3. Cleaning Area
4. Preparation zone
5. Cooking zone

Kitchen ergonomics can be applied in the following aspects

- **Kitchen Work Triangle:** The concept of kitchen work triangle plays a major role in kitchen ergonomics. The distance at which you arrange the hob, sink and fridge needs to be taken into consideration. You need to minimize walking back and forth between these areas to save energy and time. For optimum efficiency the kitchen work triangle distance needs to be kept at a minimum. The next important aspect is having enough free countertop space for food preparation.A good kitchen layout and uninterrupted work flow space improves the efficiency of kitchen.

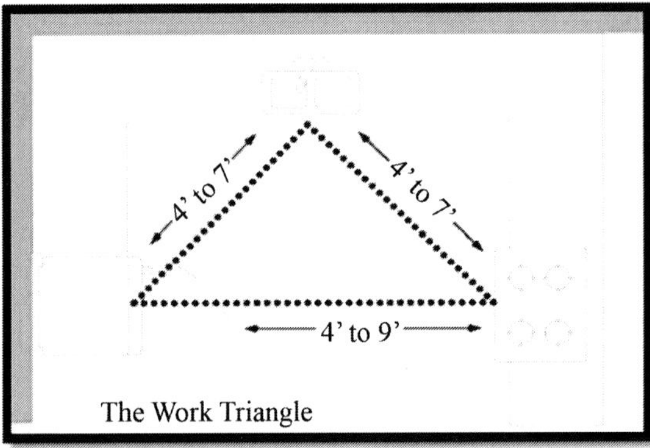

The Work Triangle

- **Kitchen Storage tips:** Storage of kitchen items is based on the daily requirement and it should be within easy reach. By segregating the items into cabinets at different levels we can avoid unnecessary bending, kneeling and stretching. This segregation highly influences the general comfort of cooking process.
    - Daily used items- closer to countertop level
    - Rarely/seldom used items-topmost or bottom row of cabinets
    - Frequently used items-should be placed in between

For base level units more drawers with full extension are better than cupboards with shelves. By doing this you can avoid a lot of bending that strains your back.

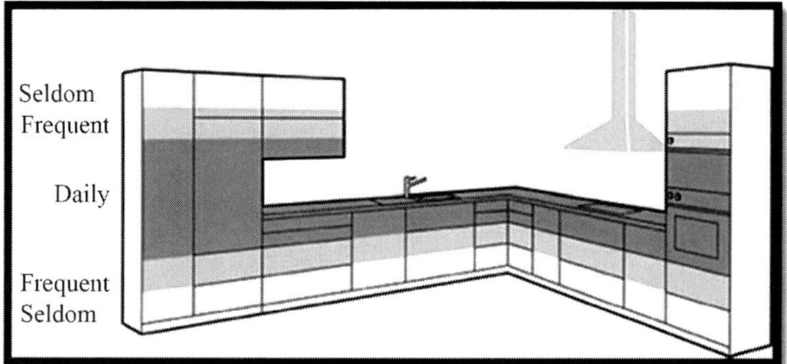

- **Working counters:** Working counters should be at a minimum height of 90 cm. If you're shorter or taller than average, you need to adapt the counter height accordingly. You can also go in for different countertop heights depending on the purpose. Be it traditional, classic or contemporary kitchen good kitchen ergonomics is the integral part. You can transform your kitchen into a comfortable working place by incorporating the right design and planning. A well thought and planned kitchen is always inviting and a marvelous place to be in.

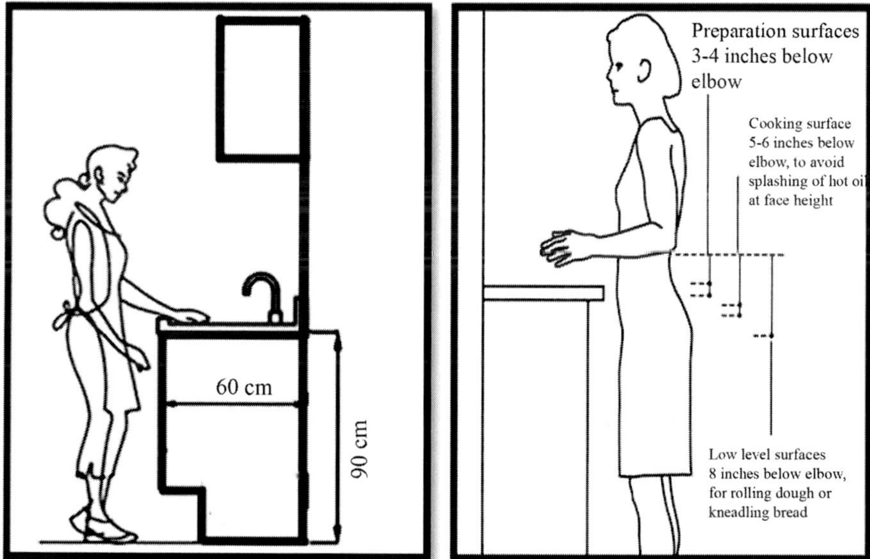

Exercise-1

Draw a kitchen design most suitable for a homemaker in today's situation taking under consideration the storage, cabinets, equipments and working counter:

Conclusion: ____

____

____

____

____

____

____

____

____

Remarks: ____

____

____

____

____

____

____

Signature

26

Market Survey

Objective: Market survey and analysis of different types of home appliances available in the market.

Exercise-1

Collect and list down the different types of home appliances with their pictures:

Name of appliance	Picture	Description

Name of appliance	Picture	Description

Conclusion:

Remarks:

Signature

References

Anshel, J. (1998): Visual Ergonomics in the Work Place, Taylor and Francis (UK, USA); published by Taylor and Francis, C., London.

Anton, T.J. (1989): Occupational Safety and Health Management, 2nd edition, (Sanfransico, Tokyo).

Bhattacharrya, A. and Meglothin (1996): Occupational Ergonomics: Theory and Applications, Marcel Dekker Inc. (New York, Basel, Hong Kong); Marcel Dekker Inc., New York.

Corlett, N. and others (1986): The Ergonomics of Working Postures: Models, Methods and Cases, Taylor and Francis (London, Philadelphia); published by Taylor and Francis, C., London.

Dul, J. and Weerdmeestre, B. (1983): Ergonomics Design for People at Work, Vol. 1., Van Nostrand Reinhold (New York), Van Nostrand Reinhold (New York).

Gottlieb, L.D. (1965): Environment and Design in Housing, Macmillan Co. New York.

Harkness, S.P. (1976): Building without Barriers for the Disabled, Whitthey Librang of Design, New York.

Heller, A. (1999): Team Workbook: Your Guide to Developing an Ergonomics Process, Lewis Publisher, Boston; Lewis Publishing Corporation.

Kroemer, K.H.E. and Grandjean, E. (1997) (5th edition): Fitting the Task to the Human, Taylor and Francis (London, New York); published by Taylor and Francis, C., London.

Kumar, S. (1999): Biomechanics in Ergonomics, Taylor and Francis (Canada); published by Taylor and Francis, C., London.

Mital A. (1988): Trends in Ergonomics/Human Factors I, Butterworths, London, Butterworths, C.

Nag, P.K. (1996): Ergonomics and Work Design: Emerging Issues in Organisational Science, New Age International, Bombay: New Age International C.

Negroponte, N. (1975): Reflections on Computer Aids to Design and Architecture, Petrocelli/ Charter, New York.

Oborne, D.J. (1982) (2nd edition): Ergonomics at Work, (Chichester, New York, Torento); Chichester: John Wiely & Sons.

Pilatowicz, G.: Eco-interiors; A Guide to Environmentally Conscious Interior Design

Pleasant, S. (2001): Body Space, Taylor and Francis (Great Britain); published by Taylor and Francis, C., London.

Pleasant, S. (1991): Ergonomics, Work and Health, Mac Millan Press, London.

Radford, A.D. and Gero, J.S. (1988): Design by Optimization in Architecture, Building and Construction, Van Nostrand Reinhold (New York).

Statham, R. (1988): House Adaptations for People, Manual for Practitioners, HMSO (London).